K. Ploog   K. Graf

# Molecular Beam Epitaxy of III-V Compounds

A Comprehensive Bibliography 1958–1983

Springer-Verlag
Berlin Heidelberg New York Tokyo 1984

Dr. Klaus Ploog
Klaus Graf

Max-Planck-Institut für Festkörperforschung, Heisenbergstraße 1
D-7000 Stuttgart 80, Fed. Rep. of Germany

ISBN 3-540-13177-9 Springer-Verlag Berlin Heidelberg New York Tokyo
ISBN 0-387-13177-9 Springer-Verlag New York Heidelberg Berlin Tokyo

This work is subject to copyright. All rights are reserved, whether the whole or part of the material is concerned, specifically those of translation, reprinting, reuse of illustrations, broadcasting, reproduction by photocopying machine or similar means, and storage in data banks. Under § 54 of the German Copyright Law where copies are made for other than private use, a fee is payable to "Verwertungsgesellschaft Wort", Munich.

© by Springer-Verlag Berlin Heidelberg 1984
Printed in Germany

The use of registered names, trademarks, etc. in this publication does not imply, even in the absence of a specific statement, that such names are exempt from the relevant protective laws and regulations and therefore free for general use.

Offset printing and bookbinding: Beltz Offsetdruck, 6944 Hemsbach/Bergstr.
2153/3130-543210

# Preface

Epitaxial growth and electronic properties of semiconductor thin films are becoming increasingly important for fundamental and applied research and for device applications. This book contains a comprehensive collection of over 1500 references covering the first 25 years of molecular beam epitaxy of III-V compound semiconductors. Molecular beam epitaxy is a versatile thin-film growth technique which emerged from the 'Three-temperature method' developed in the 1950s and from surface kinetic studies performed in the 1960s. III-V semiconductors such as GaAs, AlAs, (GaIn)As, InP, etc., play an important role in the application to optoelectronic and high-speed devices. Over the past three years the technology of molecular beam epitaxy has spread rapidly to most major research and development laboratories throughout the world, and an increasing number of highly refined III-V semiconductor structures with exactly tailored electronic properties have been produced and explored for fundamental studies as well as for device application. The comprehensive bibliography on this dramatically expanding topic helps chemists, engineers, materials scientists, and physicists working in semiconductor research and development areas to sort out the important literature of their particular interest. A direct reproduction of the output of a computer printer has been used to enable rapid publication and to keep printing costs low. The work was sponsored by the 'Bundesministerium für Forschung und Technologie' of the Federal Republic of Germany.

Stuttgart, January 1984 *K. Ploog · K. Graf*

# Contents

## Subject Categories and References

Introduction .................... *1*

Period 1958–1970 ............... *3*

Year 1971 ..................... *7*

Year 1972 ..................... *11*

Year 1973 ..................... *13*

Year 1974 ..................... *17*

Year 1975 ..................... *21*

Year 1976 ..................... *27*

Year 1977 ..................... *33*

Year 1978 ..................... *41*

Year 1979 ..................... *49*

Year 1980 ..................... *59*

Year 1981 ..................... *75*

Year 1982 ..................... *101*

Year 1983 ..................... *149*

Author Index .................. *203*

# Introduction

Molecular beam epitaxy (MBE) has now become a viable and strongly competitive thin-film growth technique in many research and development laboratories, and a variety of materials, including semiconductors, metals and insulators, have been grown as epitaxial films. MBE growth of III-V compound semiconductors and investigation of the tailored electronic properties of refined semiconductor structures have received most attention. In the past three years interest in molecular beam epitaxy has expanded dramatically, and the number of MBE-related technical papers published per year is growing exponentially (Fig. 1). This ever-increasing flood of papers makes it hard to follow up, particularly for young materials scientists newly entering this very exciting field. We have, therefore, compiled a bibliography of

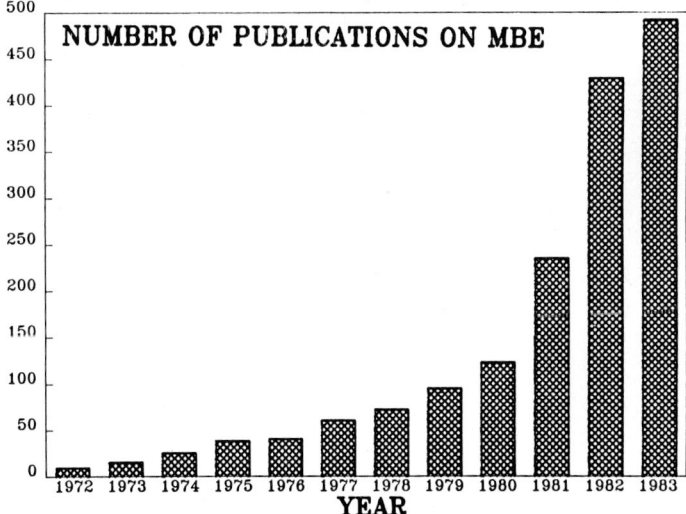

**Fig. 1.** Diagram of the number of technical papers on MBE of III-V compounds published in the period 1972 to 1983. The actual number in 1983 will increase by approximately 10 percent, because the collection of citations was finished by mid-December 1983

over 1600 references spanning the first 25 years of activities in molecular beam epitaxy of III-V compounds. References are listed with the title, using first an annual principle of classification, with the exception of early works, which are grouped together up to 1970. Within this annual classification, papers are listed alphabetically for each year by first authors in chronological succession. All citations are numbered consecutively. Citations are limited to papers that had been published up to mid-December 1983. English translations are cited if available. Review papers are marked by "R" below the number. In addition to the annual classification, the papers are subdivided into eleven subject categories that are cross-referenced. The categories - if available in the respective year - are placed in front of the alphabetic list. Each category is followed by the reference numbers of the corresponding year. The subject categories used in the bibliography are:

- AlAs, GaAs, (AlGa)As
- GaSb, InAs, Ga(AsSb), (GaIn)As
- InP, (AlIn)As, (GaIn)As, (GaIn)(PAs)
- Growth equipment
- Growth mechanism, surface analysis
- Periodic multilayer structures
- Modulation doping
- Electrical properties
- Optical properties
- Microwave devices
- Optoelectronic devices

Following the references, **all** authors with their respective reference numbers are listed alphabetically to conclude the bibliography. In addition to the experimental work a certain number of theoretical papers dealing with electronic properties of MBE-grown quantum well structures and with the quantized Hall effect are included in the bibliography and marked by "T".

# Subject Categories and References
# Period 1958–1970

**AlAs, GaAs, (AlGa)As**
  2,4,5,6,7,8,9,11,13

**GaSb, InAs, Ga(AsSb), (GaIn)As**
  1,3

**Growth Equipment**
  1,2,3,4,5,7,9

**Growth Mechanism, Surface Analysis**
  4,5,6,8,9,10,11

**Periodic Multilayer Structure**
  12,13

**Electrical Properties**
  1,2,7,13

1   Guenther,K.G.
    Aufdampfschichten aus halbleitenden III-V-Verbindungen
    Z. Naturforschg. 13a   1081-1089 (1958)

2   Davey,J.E., Pankey,T.
    Structural and optical characteristics of thin GaAs films
    J. Appl. Phys. 35   2203-2209 (1964)

3   Guenther,K.G.
R   Interfacial and condensation processes occurring with multicomponent vapours
    in "The use of thin films in physical investigations", Ed J.C. Anderson (Academic, London, 1966)   213-232 (1966)

4   Arthur,J.R.,Jr.
    Vapor pressures and phase equilibria in the Ga-As system
    J. Phys. Chem. Solids 28   2257-2267 (1967)

5   Arthur,J.R.,Jr.
    Interaction of As2, P2 and Bi molecular beams with GaAs and GaP (111) surfaces
    In: "Proc. Conf. Struct. Chem. Solid Surfaces", Ed. G.A. Somorjai (Wiley, New York, 1967)   46-1 – 46-17 (1967)

6   Arthur,J.R.,Jr.
    Interaction of Ga and As2 molecular beams with GaAs surfaces
    J. Appl. Phys. 39   4032-4034 (1968)

7   Davey,J.E., Pankey,T.
    Epitaxial GaAs films deposited by vacuum evaporation
    J. Appl. Phys. 39   1941-1948 (1968)

8   Arthur,J.R.,Jr., LePore,J.J.
    GaAs, GaP, and GaAs(x)P(1-x) epitaxial films grown by molecular beam deposition
    J. Vac. Sci. Technol. 6   545-548 (1969)

9   Cho,A.Y.
    Epitaxy by periodic annealing
    Surf. Sci. 17   494-503 (1969)

10  Cho,A.Y.
    Epitaxial growth of gallium phosphide on cleaved and polished (111) calcium fluoride
    J. Appl. Phys. 41   782-786 (1970)

11  Cho,A.Y.
    Morphology of epitaxial growth of GaAs by molecular beam method: The observation of surface structures
    J. Appl. Phys. 41   2780-2786 (1970)

**12   Esaki,L., Tsu,R.**
T    Superlattice and negative differential conductivity
       in semiconductors
     IBM J. Res. Develop. 14   61-65 (1970)

**13   Lebwohl,P.A., Tsu,R.**
T    Electrical transport properties in a superlattice
     J. Appl. Phys. 41   2664-2667 (1970)

# Subject Categories and References
# Year 1971

**AlAs, GaAs, (AlGa)As**
 14, 15, 16, 17, 18, 19, 20, 21, 22, 23, 27

**Growth Equipment**
 15, 16, 17, 20

**Growth Mechanism, Surface Analysis**
 15, 16, 18, 20

**Periodic Multilayer Structure**
 14, 22, 23, 24, 25, 26, 27

**Electrical Properties**
 15, 17, 19, 20, 23, 24, 25, 26

**Optical Properties**
 15, 17, 18, 19, 20, 21, 22, 27

**14** Cho,A.Y.
Growth of periodic structures by the molecular beam method
Appl. Phys. Lett. 19    467-468 (1971)

**15** Cho,A.Y.
Film deposition by molecular beam techniques
J. Vac. Sci. Technol. 8    S31-S38 (1971)

**16** Cho,A.Y.
GaAs epitaxy by molecular beam method: Observations of surface structure on the (001) face
J. Appl. Phys. 42    2074-2081 (1971)

**17** Cho,A.Y., Hayashi,I.
Epitaxy of silicon doped gallium arsenide by molecular beam method
Met. Trans. 2    777-780 (1971)

**18** Cho,A.Y., Hayashi,I.
Surface structures and photoluminescence of molecular beam epitaxial films of GaAs
Solid-State Electron. 14    125-132 (1971)

**19** Cho,A.Y., Hayashi,I.
P-n junction formation during molecular beam epitaxy of Ge-doped GaAs
J. Appl. Phys. 42    4422-4425 (1971)

**20** Cho,A.Y., Panish,M.B., Hayashi,I.
Molecular beam epitaxy of GaAs, $Al_xGa_{1-x}As$ and GaP
Proc. 3rd Int. Symp. on GaAs and Related Compounds (Adlard & Son, London, 1971)    18-29 (1971)

**21** Cho,A.Y., Stokowsk,S.E.
Molecular beam epitaxy and optical evaluation of $Al_xGa_{1-x}As$
Solid State Commun. 9    565-568 (1971)

**22** Kazarinov,R.F., Shmartsev,Y.V.
T  Optical phenomena due to the carriers in a semiconductor with a superlattice
Sov. Phys. Semicond. 5    710-711    *    Fiz. Tekh. Poluprov. 5    800-802 (1971)

**23** Kazarinov,R.F., Suris,R.A.
T  Possibility of the amplification of electromagnetic waves in a semiconductor with a superlattice
Sov. Phys. Semicond. 5    707-709    *    Fiz. Tekh. Poluprov. 5    797-800 (1971)

**24** Ktitorov,S.A., Simin,G.S., Sindalovskii,V.Y.
T  Bragg reflections and the high-frequency conductivity of an electronic solid-state plasma
Sov. Phys. Solid State 13 (1972)    1872-1874    *    Fiz. Tverd. Tela 13 (1971)    2230-2233 (1971)

**25  Romanov,Y.A.**
T    Periodic semiconductor structures consisting of very
       thin films
     Sov. Phys. Semicond. 5 (1972) 1256-1263   *   Fiz.
       Tekh. Poluprov. 5 (1971)    1434-1444 (1971)

**26  Stafeev,V.I.**
T    Many-layer structures consisting of a large number of
       p-n junctions
     Sov. Phys. Semicond. 5    359-365    *   Fiz. Tekh.
       Poluprov. 5    408-416 (1971)

**27  Tsu,R., Esaki,L.**
T    Nonlinear optical response of conduction electrons in
       a superlattice
     Appl. Phys. Lett. 19    246-248 (1971)

# Subject Categories and References
**Year 1972**

**AlAs, GaAs, (AlGa)As**
    28,29,32,33,34,36,37

**Growth Equipment**
    28,34

**Growth Mechanism, Surface Analysis**
    28

**Periodic Multilayer Structure**
    30,31,32,33,35,37

**Electrical Properties**
    28,30,32,33,34,36

**Optical Properties**
    30,32,35,37

28 **Cho,A.Y., Panish,M.B.**
Magnesium-doped GaAs and AlxGa1-xAs by molecular beam epitaxy
J. Appl. Phys. 43   5118-5123 (1972)

29 **Cho,A.Y., Reinhart,F.K.**
Growth of three-dimensional dielectric waveguides for integrated optics by molecular beam epitaxy
Appl. Phys. Lett. 21   355-356 (1972)

30 **Doehler,G.H.**
T  Electrical and optical properties of crystals with "nipi-superstructure"
Phys. Status Solidi B 52   533-545 (1972)

31 **Doehler,G.H.**
T  Electron states in crystals with "nipi-superstructure"
Phys. Status Solidi B 52   79-92 (1972)

32 **Esaki,L., Chang,L.L., Howard,W.E., Rideout,V.L.**
Transport properties of a GaAs-GaAlAs superlattice
Proc. 11th Int. Conf. Phys. Semicond., Warzawa - Poland, July 25-29, 1972 (Elsevier, Amsterdam, 1972)   431-437 (1972)

33 **Kazarinov,R.F., Suris,R.A.**
T  Electric and electromagnetic properties of semiconductors with a superlattice
Sov. Phys. Semicond. 6   120-131   *   Fiz. Tekh. Poluprovodn. 6   148-162 (1972)

34 **Nagashima,Y., Moriizumi,T., Takahashi,K.**
Electrical properties of vacuum deposited GaAs films
Trans. Inst. Electr. Eng. Jpn. 92 No. 2   22-28 (1972)

35 **Shik,A.Y.**
T  Optical properties of semiconductor superlattices with complex band structures
Sov. Phys. Semicond. 6 (1973)   1110-1117   *   Fiz. Tekh. Poluprovodn. 6 (1972)   1268-1277 (1972)

36 **Takahashi,K.**
Fundamental electronic transport properties of GaAs evaporated films
J. Vac. Sci. Technol. 9   502-506 (1972)

37 **Tsu,R., Jha,S.S.**
T  Phonon and polariton modes in a superlattice
Appl. Phys. Lett. 20   16-18 (1972)

# Subject Categories and References
**Year 1973**

**AlAs, GaAs, (AlGa)As**
    38,39,40,41,42,43,45,46,47,48,51,52

**Growth Equipment**
    39,40,42

**Growth Mechanism, Surface Analysis**
    38,39,40,41,42,43,45

**Periodic Multilayer Structure**
    39,40,44,46,48,49,50,52

**Electrical Properties**
    39,40,44,47,48,50,52

**Optical Properties**
    39,40,49,51

38  **Arthur, J.R., Jr.**
    The use of scanning Auger microscopy in molecular beam epitaxy of GaAs and GaP
    J. Vac. Sci. Technol. 10  136-139 (1973)

39  **Chang, L.L., Esaki, L., Howard, W.E., Ludeke, R.**
    The growth of a GaAs-GaAlAs superlattice
    J. Vac. Sci. Technol. 10  11-16 (1973)

40  **Chang, L.L., Esaki, L., Howard, W.E., Ludeke, R., Schul, G.**
    Structures grown by molecular beam epitaxy
    J. Vac. Sci. Technol. 10  655-662 (1973)

41  **Dove, D.B., Ludeke, R., Chang, L.L.**
    Interpretation of scanning high-energy electron diffraction measurements with application to GaAs surfaces
    J. Appl. Phys. 44  1897-1899 (1973)

42  **Foxon, C.T.**
  R Molecular beam epitaxy
    Acta Eletron. 16  323-329 (1973)

43  **Foxon, C.T., Harvey, J.A., Joyce, B.A.**
    The evaporation of GaAs under equilibrium and non-equilibrium conditions using a modulated beam technique
    J. Phys. Chem. Solids 34  1693-1701 (1973)

44  **Kazarinov, R.F., Suris, R.A.**
  T Theory of electrical properties of semiconductors with superlattices
    Sov. Phys. Semicond. 7  347-352  *  Fiz. Tekh. Poluprovodn. 7  488-498 (1973)

45  **Ludeke, R., Chang, L.L., Esaki, L.**
    Molecular beam epitaxy of alternating metal-semiconductor films
    Appl. Phys. Lett. 23  201-203 (1973)

46  **Mayer, J.W., Ziegler, J.F., Chang, L.L., Tsu, R., Esaki, L.**
    Profiling of periodic structures (GaAs-GaAlAs) by nuclear backscattering
    J. Appl. Phys. 44  2322-2325 (1973)

47  **Naganuma, M., Takahashi, K.**
    Epitaxial GaAs films deposited by molecular beam method
    Trans. Inst. Electr. Eng. Jpn. 93, No. 5  19-24 (1973)

48  **Price, P.J.**
  T Transport properties of the semiconductor superlattice
    IBM J. Res. Develop. 17  39-46 (1973)

**49   Romanov,Y.A., Orlov,L.K.**
T    Absorption of light in periodic semiconductor
      structures
     Sov. Phys. Semicond. 7   182-186   *   Fiz. Tekh.
        Poluprov. 7   253-260 (1973)

**50   Shik,A.Y.**
T    Transport in one-dimensional superlattices
     Sov. Phys. Semicond. 7   187-192   *   Fiz. Tekh.
        Poluprov. 7   261-269 (1973)

**51   Tracy,J.C., Wiegmann,W., Logan,R.A., Reinhart,F.K.**
     Three-dimensional light guides in single-crystal
        GaAs-Al(x)Ga(1-x)As
     Appl. Phys. Lett. 22   511-512 (1973)

**52   Tsu,R., Esaki,L.**
T    Tunneling in a finite superlattice
     Appl. Phys. Lett. 22   562-564 (1973)

# Subject Categories and References
# Year 1974

**AlAs, GaAs, (AlGa)As**
53,54,55,56,57,58,59,60,61,62,63,64,65,66,68,69,70,71,
72,73,74,75,76,77

**InP, (AlIn)As, (GaIn)As, (GaIn)(PAs)**
67,69

**Growth Equipment**
53,56,57,59,61,65,68,71

**Growth Mechanism, Surface Analysis**
53,59,61,65,67,68,69,70,71,73,75

**Periodic Multilayer Structure**
54,60,61,62,63,64,65,66,73,76,77

**Electrical Properties**
54,55,57,58,59,61,62,63,64,65,66,74,76,77

**Optical Properties**
56,60,76

**Microwave Devices**
58,59

**Optoelectronic Devices**
56

53 **Arthur,J.R.,Jr.**
Surface stoichiometry and structure of GaAs
Surf. Sci. 43    449-461 (1974)

54 **Chang,L.L., Esaki,L., Segmueller,A., Tsu,R.**
Resonant electron transport in semiconductor barrier structures
Proc. 12th Int. Conf. Phys. Semicond., Stuttgart, Germany, July 15-19, 1974 (Teubner, Stuttgart, 1974)    688-692 (1974)

55 **Chang,L.L., Esaki,L., Tsu,R.**
Resonant tunneling in semiconductor double barriers
Appl. Phys. Lett. 24    593-595 (1974)

56 **Cho,A.Y., Casey,H.C.,Jr.**
GaAs-$Al_xGa_{1-x}As$ double-heterostructure laser prepared by molecular beam epitaxy
Appl. Phys. Lett. 35    288-290 (1974)

57 **Cho,A.Y., Casey,H.C.,Jr.**
Properties of Schottky barriers and p-n junctions prepared with GaAs and $Al_xGa_{1-x}As$ molecular epitaxial layers
J. Appl. Phys. 45    1258-1263 (1974)

58 **Cho,A.Y., Dunn,C.N., Kuvas,R.L., Schroeder,W.E.**
GaAs IMPATT diodes prepared by molecular beam epitaxy
Appl. Phys. Lett. 25    224-226 (1974)

59 **Cho,A.Y., Reinhart,F.K.**
Interface and doping profile characteristics with molecular beam epitaxy of GaAs: GaAs voltage varactor
J. Appl. Phys. 45    1812-1817 (1974)

60 **Dingle,R., Wiegmann,W., Henry,C.H.**
Quantum states of confined carriers in very thin $Al_xGa_{1-x}As$-GaAs-$Al_xGa_{1-x}As$ heterostructures
Phys. Rev. Lett. 33    827-830 (1974)

61 **Esaki,L.**
R   Computer-controlled molecular beam epitaxy
Proc. 6th Int. Vacuum Congr., Jpn. J. Appl. Phys. Suppl. 2, Pt. 1    821-828 (1974)

62 **Esaki,L.**
R   Long journey into tunneling
Rev. Mod. Phys. 46    237-244 (1974)

63 **Esaki,L.**
R   Long journey into tunneling
Proc. IEEE 62    825-831 (1974)

64   Esaki,L.
  R  Long journey into tunneling
     Science 183   1149-1155 (1974)

65   Esaki,L.
     Computer-controlled molecular beam epitaxy
     Buturi 43   452-460   [J. Jpn. Soc. Appl. Phys. 43,
       Suppl.   452-460] (1974)

66   Esaki,L., Chang,L.L.
     New transport phenomenon in a semiconductor
       "superlattice"
     Phys. Rev. Lett. 33   495-498 (1974)

67   Farrow,R.F.C.
     Growth of indium phosphide films from In and P2 beams
       in ultra-high vacuum
     J. Phys. D 7   L121-L124 (1974)

68   Foxon,C.T., Boudry,M.R., Joyce,B.A.
     Evaluation of surface kinetic data by the transform
       analysis of modulated beam measurements
     Surf. Sci. 44   69-92 (1974)

69   Foxon,C.T., Joyce,B.A., Farrow,R.F.C.,
       Griffiths,R.M.
     The identification of species evolved in the
       evaporation of III-V compounds
     J. Phys. D 7   2422-2435 (1974)

70   Hottmann,H., Schulz,M.
     Reflection splitting in RHEED investigations of
       vacuum evaporated GaAs layers
     Krist. Technik 9   661-673 (1974)

71   Joyce,B.A.
  R  The growth and structure of semiconducting thin films
     Rep. Prog. Phys. 37   363-420 (1974)

72   Ludeke,R., Esaki,L.
     Electron energy-loss spectroscopy of GaAs and Ge
       surfaces
     Phys. Rev. Lett. 33   653-656 (1974)

73   Ludeke,R., Esaki,L., Chang,L.L.
     Ga(1-x)Al(x)As superlattices profiled by Auger
       electron spectroscopy
     Appl. Phys. Lett. 24   417-419 (1974)

74   Naganuma,M., Takahashi,K.
     Impurity doping effect of molecular-beam epitaxial
       GaAs films
     Trans. Inst. Electr. Eng. Jpn. 94, No. 5   29-36
       (1974)

75 **Rosin,H., Schulz,M.**
Die Praeparation von GaAs-Substraten, untersucht mit
Reflexionselektroneninterferenzen
Krist. Technik 9   905-912 (1974)

76 **Shik,A.Y.**
R  Superlattices - periodic semiconductor structures
Sov. Phys. Semicond. 8 (1975)   1195-1209   *   Fiz.
Tekh. Poluprov. 8 (1974)   1841-1864 (1974)

77 **Tsu,R., Janak,J.**
T  Magnetic quantization in a superlattice
Phys. Rev. B 9   404-408 (1974)

# Subject Categories and References
**Year 1975**

**AlAs, GaAs, (AlGa)As**
  78,79,80,81,82,83,84,85,86,87,88,89,90,91,92,94,95,96,
  97,99,100,101,102,103,104,105,106,107,108,109,110,111,
  112,113,114

**Growth Equipment**
  78,80,83,85,86,91,95,97,98,99,105,107,108,109

**Growth Mechanism, Surface Analysis**
  80,83,84,90,91,97,107,108

**Periodic Multilayer Structure**
  85,86,87,89,92,93,103,104,110,111,112,114

**Electrical Properties**
  80,81,82,83,84,85,89,93,95,96,101,104,106,107,108,110,
  111

**Optical Properties**
  78,79,82,83,85,86,87,88,91,92,93,94,95,96,102,103,104,
  110,112,113,114,115

**Microwave Devices**
  82,83,106

**Optoelectronic Devices**
  78,79,82,83,94,113

**78** Casey,H.C.,Jr., Cho,A.Y., Barnes,P.A.
Application of molecular-beam epitaxial layers to heterostructure lasers
IEEE J. Quantum Electron. QE-11  467-470 (1975)

**79** Casey,H.C.,Jr., Somekh,S., Ilegems,M.
Room-temperature operation of low-threshold separate-confinement heterostructure injection laser with distributed feedback
Appl. Phys. Lett. 27  142-144 (1975)

**80** Chang,L.L., Ludeke,R.
R Molecular beam epitaxy
In: "Epitaxial Growth", Ed. J.W. Matthews (Academic Press, New York, 1975)  37-72 (1975)

**81** Cho,A.Y.
Impurity profiles of GaAs epitaxial layers doped with Sn,Si,and Ge grown with molecular beam epitaxy
J. Appl. Phys. 46  1733-1735 (1975)

**82** Cho,A.Y.
Device fabrication by molecular-beam epitaxy
IEDM 75 Technical Digest  429-432 (1975)

**83** Cho,A.Y., Arthur,J.R.,Jr.
R Molecular beam epitaxy
Progr. Solid State Chem. 10  157-191 (1975)

**84** Cho,A.Y., Ballamy,W.C.
GaAs planar technology by molecular beam epitaxy (MBE)
J. Appl. Phys. 46  783-785 (1975)

**85** Dingle,R.
R Confined carrier quantum states in ultrathin semiconductor heterostructures
in 'Festkoerperprobleme', Ed. H.J.Queisser(Pergamon-Vieweg, Braunschweig,1975)
Vol. XV  21-48 (1975)

**86** Dingle,R.
R Optical and electronic properties of thin $Al(x)Ga(1-x)As/GaAs$ heterostructures
CRC Crit. Rev. Solid State & Mater. Sci. 5  585-590 (1975)

**87** Dingle,R., Gossard,A.C., Wiegmann,W.
Direct observation of superlattice formation in a semiconductor heterostructure
Phys. Rev. Lett. 34  1327-1330 (1975)

**88** Dingle,R., Wiegmann,W.
Optical investigation of stress in the central GaAs layer of molecular-beam-grown $Al_xGa_{1-x}As$-GaAs-$Al_xGa_{1-x}As$ structures
J. Appl. Phys. 46  4312-4315 (1975)

89  Doehler,G.H., Tsu,R., Esaki,L.
T   A new mechanism for negative differential
       conductivity in superlattices
    Solid State Commun. 17   317-320  (1975)

90  Foxon,C.T., Joyce,B.A.
    Interaction kinetics of As4 and Ga on (100) GaAs
       surfaces using a modulated molecular beam technique
    Surf. Sci. 50   434-450  (1975)

91  Gonda,S., Matsushima,Y., Makita,Y., Mukai,S.
    Characterization and substrate-temperature dependence
       of crystalline state of GaAs grown by molecular
       beam epitaxy
    Jpn. J. Appl. Phys. 14   935-942  (1975)

92  Gover,A., Yariv,A.
T   Intraband radiative transitions and
       plasma-electromagnetic-wave coupling in periodic
       semiconductor structure
    J. Appl. Phys. 46   3946-3950  (1975)

93  Ignatov,A.A., Romanov,Y.A.
T   Self-induced transparency in semiconductors with
       superlattices
    Sov. Phys. Solid State 17   2216-2217   *   Fiz.
       Tverd. Tela 17   3388-3389  (1975)

94  Ilegems,M., Casey,H.C.,Jr., Somekh,S., Panish,M.B.
    Epitaxial growth on optical gratings for distributed
       feedback GaAs injection lasers
    J. Cryst. Growth 31   158-164  (1975)

95  Ilegems,M., Dingle,R.
    Acceptor incorporation in GaAs grown by beam epitaxy
    Inst. Phys. Conf. Ser. 24   1-9  (1975)

96  Ilegems,M., Dingle,R., Rupp,L.W.
    Optical and electrical properties of Mn-doped GaAs
       grown by molecular beam epitaxy
    J. Appl. Phys. 46   3059-3065  (1975)

97  Joyce,B.A., Foxon,C.T.
R   Kinetic studies of the growth of III-V compounds
       using modulated molecular beam techniques
    J. Cryst. Growth 31   122-129  (1975)

98  Kakati,D., Sharma,B.D.
T   AB3-type molecular source for beam epitaxy
    J. Vac. Sci. Technol. 12   956-957  (1975)

99  Kometani,T.Y., Wiegmann,W.
    Measurement of Ga and Al in a molecular beam epitaxy
       chamber by atomic absorption spectrometry (AAS)
    J. Vac. Sci. Technol. 12   933-936  (1975)

100 **Ludeke,R., Esaki,L.**
Electron spectroscopy of GaAs and AlAs surfaces
Surf. Sci. 47   132-142 (1975)

101 **Matsunaga,N., Naganuma,M., Takahashi,K.**
R   Molecular beam epitaxy with ionised dopant
Trans. Inst. Electr. Eng. Jpn. 95 No. 12   23-28 (1975)

102 **Merz,J.L., Logan,R.A., Wiegmann,W., Gossard,A.C.**
Taper couplers for GaAs-Al$_x$Ga$_{1-x}$As waveguide layers produced by liquid phase and molecular beam epitaxy
Appl. Phys. Lett. 26   337-340 (1975)

103 **Mukherji,D., Nag,B.R.**
T   Band structure of semiconductor superlattices
Phys. Rev. B 12   4338-4345 (1975)

104 **Mukherji,D., Nag,B.R.**
T   Miniband parameters of semiconductor superlattices
Solid-State Electron. 18   1107-1109 (1975)

105 **Murayama,Y.**
Structures of GaAs and GaP thin films formed by rf ion-beam epitaxy
J. Vac. Sci. Technol. 12   876-878 (1975)

106 **Naganuma,M., Kamimura,K., Takahashi,K., Sakai,Y.**
GaAs FET prepared with molecular beam epitaxial films
Jpn. J. Appl. Phys. 14   581-582 (1975)

107 **Naganuma,M., Takahashi,K.**
GaAs, GaP, and GaAs$(1-x)$P$(x)$ films deposited by molecular beam epitaxy
Phys. Status Solidi A 31   187-200 (1975)

108 **Naganuma,M., Takahashi,K.**
Ionized Zn doping of GaAs molecular beam epitaxial films
Appl. Phys. Lett. 27   342-344 (1975)

109 **Tateishi,K., Naganuma,M., Takahashi,K.**
Molecular beam epitaxial GaP, Ga(AsP)
Trans. Inst. Electr. Eng. Jpn. 95, No. 3   11-19 (1975)

110 **Tsu,R., Chang,L.L., Sai-Halasz,G.A., Esaki,L.**
Effects of quantum states on the photocurrent in a "superlattice"
Phys. Rev. Lett. 34   1509-1512 (1975)

111 **Tsu,R., Doehler,G.H.**
T   Hopping conduction in a "superlattice"
Phys. Rev. B 12   680-686 (1975)

112  Tsu,R., Koma,A., Esaki,L.
     Optical properties of semiconductor superlattice
     J. Appl. Phys. 46   842-843 (1975)

113  Ziel,J.P. van der, Dingle,R., Miller,R.C.,
         Wiegmann,W., Nordland,W.A.
     Laser oscillation from quantum states in very thin
         GaAs-Al0.2Ga0.8As multilayer structures
     Appl. Phys. Lett. 26   463-465 (1975)

114  Ziel,J.P. van der, Ilegems,M.
     Multilayer GaAs-Al0.3Ga0.7As dielectric quarter wave
         stacks grown by molecular beam epitaxy
     Appl. Optics 14   2627-2630 (1975)

115  Ziel,J.P. van der, Mikulyak,R.M., Cho,A.Y.
     Second harmonic generation in a GaP waveguide
     Appl. Phys. Lett. 27   71-73 (1975)

# Subject Categories and References
# Year 1976

**AlAs, GaAs, (AlGa)As**
   116,117,118,119,120,121,122,124,125,126,127,129,130,131,
   132,133,134,135,136,137,140,141,142,143,146,148,150,151,
   152,153,154,155

**GaSb, InAs, Ga(AsSb), (GaIn)As**
   147

**InP, (AlIn)As, (GaIn)As, (GaIn)(PAs)**
   139,147

**Growth Equipment**
   122,124,125,126,127,137,139,147,149,150

**Growth Mechanism, Surface Analysis**
   119,124,126,127,128,132,135,137,138,139,147,149,150

**Periodic Multilayer Structure**
   118,122,123,124,125,129,131,136,141,142,143,144,145,146,
   148,152,153,154

**Electrical Properties**
   116,120,124,125,130,131,133,139,142,143,144,145,146,151,
   155

**Optical Properties**
   121,122,123,124,129,134,136,138,140,141,142,148,152,153,
   154

**Microwave Devices**
   116,120,130,151,155

**Optoelectronic Devices**
   121,123,134,141

116 **Ballamy,W.C., Cho,A.Y.**
Planar isolated GaAs devices produced by molecular
beam epitaxy
IEEE Trans. Electron Devices ED-23    481-484 (1976)

117 **Chang,L.L., Koma,A.**
Interdiffusion between GaAs and AlAs
Appl. Phys. Lett. 29    138-141 (1976)

118 **Chang,L.L., Segmueller,A., Esaki,L.**
Smooth and coherent layers of GaAs and AlAs grown by
molecular beam epitaxy
Appl. Phys. Lett. 28    39-41 (1976)

119 **Cho,A.Y.**
Bonding direction and surface-structure orientation
on GaAs (001)
J. Appl. Phys. 47    2841-2843 (1976)

120 **Cho,A.Y., Ch'en,D.R.**
GaAs MESFET prepared by molecular beam epitaxy (MBE)
Appl. Phys. Lett. 28    30-31 (1976)

121 **Cho,A.Y., Dixon,R.W., Casey,H.C.,Jr., Hartman,R.L.**
Continuous room-temperature operation of
GaAs-Al$_x$Ga$_{1-x}$As double-heterostructure lasers
prepared by molecular-beam epitaxy
Appl. Phys. Lett. 28    501-503 (1976)

122 **Dingle,R.**
R   Optical properties of semiconductor superlattices
Proc. 13th Int. Conf. Phys. Semicond., Rome-Italy,
Aug. 30 - Sept. 3, 1976 (North-Holland, Amsterdam,
1976)    965-974 (1976)

123 **Elachi,C.**
TR  Waves in active and passive periodic structures: a
review
Proc. IEEE 64    1666-1698 (1976)

124 **Esaki,L., Chang,L.L.**
R   Semiconductor superfine structures by
computer-controlled molecular beam epitaxy
Thin Solid Films 36    285-298 (1976)

125 **Esaki,L., Chang,L.L.**
R   Superfine structures of semiconductors grown by
molecular beam epitaxy
CRC Crit. Rev. Solid State & Mater. Sci. 6    195-208
(1976)

126 **Foxon,C.T., Joyce,B.A., Holloway,S.**
Instrument response function of a quadrupole mass
spectrometer used in time-of-flight measurements
Int. J. Mass Spectr. Ion Phys. 21    241-255 (1976)

127  Geng,P., Ranke,W., Jacobi,K.
A multiple technique UHV chamber for the investigation of epitaxially grown semiconductor surfaces
J. Phys. E 9   924-925 (1976)

128  Gonda,S., Matsushima,Y.
Effect of substrate temperature on composition ratio x in molecular beam epitaxial GaAs(1-x)P(x)
J. Appl. Phys. 47   4198-4200 (1976)

129  Gossard,A.C., Petroff,P.M., Wiegmann,W., Dingle,R., Savage,A.
Epitaxial structures with alternate-atomic-layer composition modulation
Appl. Phys. Lett. 29   323-325 (1976)

130  Hollan,L.
R   Evolution of the techniques of preparation of epitaxial GaAs for hyperfrequency devices
Vide 31   131-137 (1976)

131  Ignatov,A.A., Romanov,Y.A.
T   Nonlinear electromagnetic properties of semiconductors with a superlattice
Phys. Status Solidi B 73   327-333 (1976)

132  Jewsbury,P., Holloway,S.
T   The interaction of As4 and Ga beams on a GaAs(100) surface
J. Phys. C 9   3205-3215 (1976)

133  Lang,D.V., Cho,A.Y., Gossard,A.C., Ilegems,M., Wiegmann,W.
Study of electron traps in n-GaAs grown by molecular beam epitaxy
J. Appl. Phys. 47   2558-2564 (1976)

134  Lee,T.P., Cho,A.Y.
Single-transverse-mode injection lasers with embedded stripe layer grown by molecular beam epitaxy
Appl. Phys. Lett. 29   164-166 (1976)

135  Ludeke,R., Koma,A.
Electronic surface states on clean and oxygen-exposed GaAs surfaces
J. Vac. Sci. Technol. 13   241-247 (1976)

136  Manuel,P., Sai-Halasz,G.A., Chang,L.L., Chang,C.A., Esaki,L.
Resonant Raman scattering in a semiconductor superlattice
Phys. Rev. Lett. 37   1701-1704 (1976)

137 **Massies,J., Etienne,P., Linh,N.T.**
Epitaxie par jets moléculaires
Rev. Tech. Thomson-CSF 8  5-39 (1976)

138 **Matsushima,Y., Gonda,S.**
Molecular beam epitaxy of GaP and GaAs(1-x)P(x)
Jpn. J. Appl. Phys. 15  2093-2101 (1976)

139 **Matsushima,Y., Hirofuji,Y., Gonda,S., Mukai,S., Kimata,M.**
Molecular beam epitaxial growth of InP
Jpn. J. Appl. Phys. 15  2321-2325 (1976)

140 **Merz,J.L., Cho,A.Y.**
Low-loss AlxGa1-xAs waveguides grown by molecular beam epitaxy
Appl. Phys. Lett. 28  456-458 (1976)

141 **Miller,R.C., Dingle,R., Gossard,A.C., Logan,R.A., Nordland,W.A., Wiegmann,W.**
Laser oscillation with optically pumped very thin GaAs-AlxGa1-xAs multilayer structures and conventional double heterostructures
J. Appl. Phys. 47  4509-4517 (1976)

142 **Mukherji,D., Nag,B.R.**
T Electron effective mass in semiconductor superlattices
Phys. Status Solidi B 75  K35-K37 (1976)

143 **Mukherji,D., Nag,B.R.**
T On the density of states in a semiconductor superlattice in the presence of a quantizing magnetic field
Int. J. Electron. 41  263-272 (1976)

144 **Pavlovich,V.V., Epshtein,E.M.**
T Nonlinear high-frequency conductivity of superlattices
Sov. Phys. Solid State 19  863-864  *  Fiz. Tverd. Tela 18  1483-1485 (1976)

145 **Pavlovich,V.V., Epshtein,E.M.**
T Conductivity of a superlattice semiconductor in strong electric fields
Sov. Phys. Semicond. 10  1196-1197  *  Fiz. Tekh. Poluprovodn. 10  2001-2003 (1976)

146 **Shmelev,G.M., Chaikovskii,I.A., Shon,C.M.**
T The conductivity of semiconductors with superlattice
Phys. Status Solidi B 76  811-816 (1976)

147 **Tateishi,K., Naganuma,M., Takahashi,K.**
Graded-bandgap III-V ternary compound films by molecular beam epitaxy
Jpn. J. Appl. Phys. 15  785-789 (1976)

148 **Tsu,R., Esaki,L.**
T    Raman scattering for lo phonons via quantum states
     Proc. 3rd Int. Conf. Light Scattering Solids, Eds. M
        Balkanski, R.C.C. Leite, S.P.S. Porto (Flammarion
        Sci., Paris, 1976)    533-538 (1976)

149  **Venables,J.A.**
     An UHV SEM for in-situ deposition and surface studies
     in "Developments in Electron Microscopy and
        Analysis", Ed. J.A.Venables (Academic Press,
        London, 1976)    23-26 (1976)

150  **Venables,J.A., Janssen,A.P., Harland,C.J.,
        Joyce,B.A.**
     Scanning Auger electron microscopy at 30 nm resolution
     Phil. Mag. 34    495-500 (1976)

151  **Wood,C.E.C.**
     Molecular beam epitaxial GaAs layers for MESFET's
     Appl. Phys. Lett. 29    746-748 (1976)

152  **Ziel,J.P. van der, Ilegems,M.**
     Interference filters: single crystal multilayer
        AlAs-GaAs
     Appl. Optics 15    1256-1257 (1976)

153  **Ziel,J.P. van der, Ilegems,M.**
     Optical second harmonic generation in periodic
        multilayer GaAs-Al0.3Ga0.7As structures
     Appl. Phys. Lett. 28    437-439 (1976)

154  **Ziel,J.P. van der, Ilegems,M., Mikulyak,R.M.**
     Optical birefringence of thin GaAs-AlAs multilayer
        films
     Appl. Phys. Lett. 28    735-737 (1976)

155  **Zuleeg,R.**
     Developments in GaAs FETs and ICs
     IEDM 76 Technical Digest    347-350 (1976)

# Subject Categories and References
**Year 1977**

**AlAs, GaAs, (AlGa)As**
157,158,159,160,164,165,167,168,169,171,172,173,174,175,
177,179,180,181,182,183,184,186,187,189,190,191,192,194,
196,197,198,199,200,203,204,205,206,207,209,210,212,214,
215

**GaSb, InAs, Ga(AsSb), (GaIn)As**
162,163,166,171,178,201,202,211,213

**InP, (AlIn)As, (GaIn)As, (GaIn)(PAs)**
156,162,163,170,171,178,188,200,201,202

**Growth Equipment**
171,173,174,179,181,184,187,188,192,194,197,205,211,213

**Growth Mechanism, Surface Analysis**
162,170,171,174,175,180,181,182,183,186,187,188,191,192,
197,198,202,209,210,211,213

**Periodic Multilayer Structure**
157,159,160,161,164,165,169,171,172,177,184,185,189,190,
193,194,195,196,201,204,206,208,209,214,215

**Electrical Properties**
156,158,161,162,163,164,165,167,171,172,176,177,178,179,
181,182,186,187,188,193,194,202,203,212,213

**Optical Properties**
156,159,160,165,166,169,171,177,178,179,188,189,190,194,
195,199,204,205,207,208,212,214,215

**Microwave Devices**
163,165,167,174,181,184,203

**Optoelectronic Devices**
165,174,181,184,199,205

156 **Bachmann,K.J., Buehler,E., Miller,B.I., McFee,J.H., Thiel,F.A.**
R  The current status of the preparation of single crystals, bicrystals, and epitaxial layers of p-InP and of polycrystalline p-InP films for photovoltaic applications
J. Cryst. Growth 39   137-150 (1977)

157 **Balibar,F.**
R  L'epitaxie par jets moleculaires
Recherche 8 No. 83   984-987 (1977)

158 **Baraff,G.A., Appelbaum,J.A., Hamann,D.R.**
T  Electronic structure at an abrupt GaAs-Ge interface
J. Vac. Sci. Technol. 14   999-1005 (1977)

159 **Caruthers,E., Lin-Chung,P.J.**
T  Electronic structure of GaAs-Ga(1-x)Al(x)As repeated monolayer heterostructure
Phys. Rev. Lett. 26   1543-1546 (1977)

160 **Caruthers,E., Lin-Chung,P.J.**
T  Pseudopotential calculations for $(GaAs)_1-(AlAs)_1$ and related monolayer heterostructures
Phys. Rev. B 17   2705-2717 (1977)

161 **Chaikovskii,I.A., Shmelev,G.M., Hung,C.Q.**
T  Magneto-oscillatory effects in semiconductors with superlattice
J. Phys. C 10   3315-3322 (1977)

162 **Chang,C.A., Ludeke,R., Chang,L.L., Esaki,L.**
   Molecular-beam epitaxy (MBE) of $In(1-x)Ga(x)As$ and $GaSb(1-y)As(y)$
Appl. Phys. Lett. 31   759-761 (1977)

163 **Chang,L.L., Esaki,L.**
T  Tunnel triode - a tunneling base transistor
Appl. Phys. Lett. 31   687-689 (1977)

164 **Chang,L.L., Sakaki,H., Chang,C.A., Esaki,L.**
   Shubnikov-de Haas oscillations in a semiconductor superlattice
Phys. Rev. Lett. 38   1489-1493 (1977)

165 **Cho,A.Y.**
R  Preparation and properties of GaAs devices by molecular beam epitaxy
Jpn. J. Appl. Phys. 16   Suppl. 16-1,435-442 (1977)

166 **Cho,A.Y., Casey,H.C.,Jr., Foy,P.W.**
   Back-surface emitting $GaAs_xSb_{1-x}$LED's prepared by molecular beam epitaxy
Appl. Phys. Lett. 30   397-399 (1977)

167  Cho,A.Y., DiLorenzo,J.V., Hewitt,B.S.,
        Niehaus,W.C., Schlosser,W.O., Radice,C.
     Low-noise and high-power GaAs mircowave field-effect
        transistor prepared by molecular beam epitaxy
     J. Appl. Phys. 48   346-349 (1977)

168  Cho,A.Y., DiLorenzo,J.V., Mahoney,G.E.
     Selective lift-off for preferential growth with
        molecular beam epitaxy
     IEEE Trans. Electron devices ED-24   1186-1187 (1977)

169  Cho,A.Y., Yariv,A., Yeh,P.
     Observation of confined propagation in Bragg
        waveguides
     Appl. Phys. Lett. 30   471-472 (1977)

170  Dowsett,M.G., King,R.M., Parker,E.H.C.
     SIMS evaluation of contamination on ion-cleaned(100)
        InP substrates
     Appl. Phys. Lett. 31   529-531 (1977)

171  Esaki,L.
  R  Exploratory investigation for quantum effects in
        semiconductor heterostructures
     Proc. 7th Intern. Vac. Congr. & 3rd Intern. Conf.
        Solid Surfaces (Vienna 1977)   1907-1914 (1977)

172  Esaki,L., Chang,L.L.
  R  Superfine structures of semiconductors grown by
        molecular beam epitaxy
     Chem. Phys. Solid Surf.   111-124 (1977)

173  Etienne,P., Massies,J., Linh,N.T.
     Un systeme de jets moleculaires asservis a un
        spectometre de masse quadrupolaire
     J. Phys. E 10   1153-1155 (1977)

174  Farrow,R.F.C.
  R  Molecular beam epitaxy
     In: "1976 Crystal Growth and Materials", Eds.
        E.Kaldis and H.J.Scheel (North-Holland, Amsterdam,
        1977) p. 235-277 (1977)

175  Foxon,C.T., Joyce,B.A.
     Interaction kinetics of As2 and Ga on (100) GaAs
        surfaces
     Surf. Sci. 64   293-304 (1977)

176  Harrison,W.A.
  T  Elementary theory of heterojunctions
     J. Vac. Sci. Technol. 14   1016-1021 (1977)

177  Herrick,D.R.
  T  Construction of bound states in the continuum for
        epitaxial heterostructure superlattices
     Physica 85 B   44-50 (1977)

178  **Hiyamizu,S., Fujii,T., Nanbu,K., Maekawa,S.**
Properties of heteroepitaxial In(x)Ga(1-x)As by
molecular beam epitaxy
Jpn. J. Appl. Phys. 17, Suppl. 17-1   79-85 (1977)

179  **Ilegems,M.**
Beryllium doping and diffusion in molecular beam
epitaxy of GaAs and AlxGa1-xAs
J. Appl. Phys. 48   1278-1287 (1977)

180  **Janssen,A.P., Venables,J.A.**
Surface physics experiments in an U.H.V. scanning
electron microscope
Inst. Phys. Conf. Ser. 36   91-94 (1977)

181  **Joyce,B.A., Foxon,C.T.**
R  Growth and doping of semiconductor films by molecular
beam epitaxy
Inst. Phys. Conf. Ser. 32   17-37 (1977)

182  **Joyce,B.A., Foxon,C.T.**
Growth and doping kinetics in molecular beam epitaxy
Jpn. J. Appl. Phys. 16, Suppl. 16-1   17-23 (1977)

183  **Laurence,G., Samuel,G.S., Joyce,B.A., Foxon,C.T.,
Janssen,G., Venables,J.A.**
Adsoption-desorption studies of Zn on GaAs
Surf. Sci. 68   190-203 (1977)

184  **Luscher,P.E.**
R  Crystal growth by molecular beam epitaxy
Solid State Technol. 20 No. 12   43-52 (1977)

185  **Maslov,V.N.**
R  Semiconducting periodic structures: a new class of
semiconducting materials
Inorg. Mater. 13   923-930   *   Izvest. Akad. Nauk
SSSR, Neorg. Mater. 13   1133-1142 (1977)

186  **Massies,J., Devoldere,P., Etienne,P., Linh,N.T.**
Applications of molecular beam epitaxy to the study
of surface properties of III-V compounds
Proc. 7th Int. Vac. Congr. & 3rd Int. Conf. Solid
Surfaces (Vienna 1977) p 639-641 (1977)

187  **Matsunaga,N., Naganuma,M., Takahashi,K.**
Molecular beam epitaxy with ionized beam doping
Jpn. J. Appl. Phys. 16 Suppl. 16-1   443-449 (1977)

188  **McFee,J.H., Miller,B.I., Bachmann,K.J.**
Molecular beam epitaxial growth of InP
J. Electrochem. Soc. 124   261-272 (1977)

189  Merz,J.L., Barker,A.S.,Jr., Gossard,A.C.
     Raman scattering and zone-folding effects for
        alternating monolayers of GaAs-AlAs
     Appl. Phys. Lett. 31    117-119 (1977)

190  Merz,J.L., Gossard,A.C., Wiegmann,W.
     Alternate-monolayer single-crystal GaAs-AlAs optical
        waveguides
     Appl. Phys. Lett. 30    629-631 (1977)

191  Nagata,S., Tanaka,T.
     Self-masking selective epitaxy by molecular-beam
        method
     J. Appl. Phys. 48    940-942 (1977)

192  Nagata,S., Tanaka,T., Fukai,M.
     Self-aligned three-dimensional Ga(1-x)Al(x)As
        structures grown by molecular beam epitaxy
     Appl. Phys. Lett. 30    503-505 (1977)

193  Orlov,L.K., Romanov,Y.A.
  T  Nonlinear interaction of two waves in semiconductors
        with a superlattice
     Sov. Phys. Solid State 19    421-424    *    Fiz. Tverd.
        Tela 19    726-731 (1977)

194  Panish,M.B.
  R  Molecular beam epitaxy: new technique for growing
        crystals
     Bell Labs. Record 55    109-114 (1977)

195  Pavlovich,V.V., Epshtein,E.M.
  T  Quantum theory of electromagnetic wave absorption by
        free carriers in a semiconductor with a
        superlattice
     Sov. Phys. Solid State 19    1616-1618    *    Fiz.
        Tverd. Tela 19    2760-2764 (1977)

196  Petroff,P.M.
  R  Transmission electron microscopy of interfaces in
        III-V compound semiconductors
     J. Vac. Sci. Technol. 14    973-978 (1977)

197  Ploog,K., Fischer,A.
     In situ characterization of MBE grown GaAs and
        Al(x)Ga(1-x)As films using RHEED, SIMS, and AES
        techniques
     Appl. Phys. 13    111-121 (1977)

198  Ploog,K., Fischer,A., Raisch,F.
     Molecular beam epitaxy of GaAs and simultaneous
        characterization by RHEED, SIMS, and AES techniques
     Proc. 7th Intern. Vac. Congr. & 3rd Intern. Conf.
        Solid Surfaces (Vienna 1977)    1705-1708 (1977)

199 **Reinhart,F.K., Cho,A.Y.**
AlyGa1-yAs-AlxGa1-xAs laser structures for integrated
optics grown by molecular beam epitaxy
Appl. Phys. Lett. 31    457-459 (1977)

200 **Rode,D.L., Wagner,W.R., Schumaker,N.E.**
Singular instabilities on LPE GaAs, CVD Si, and MBE
InP growth surfaces
Appl. Phys. Lett. 30    75-78 (1977)

201 **Sai-Halasz,G.A., Tsu,R., Esaki,L.**
T A new semiconductor superlattice
Appl. Phys. Lett. 30    651-653 (1977)

202 **Sakaki,H., Chang,L.L., Ludeke,R., Chang,C.A., Sai-Halasz,G.A., Esaki,L.**
In(1-x)Ga(x)As-GaSb(1-y)As(y) heterojunctions by
molecular beam epitaxy
Appl. Phys. Lett. 31    211-213 (1977)

203 **Schneider,M.V., Linke,R.A., Cho,A.Y.**
Low-noise millimeter-wave mixer diodes prepared by
molecular beam epitaxy (MBE)
Appl. Phys. Lett. 31    219-221 (1977)

204 **Schulman,J.N., McGill,T.C.**
T Band structure of AlAs-GaAs(100) superlattices
Phys. Rev. Lett. 39    1680-1683 (1977)

205 **Scifres,D.R., Burnham,R.D., Streifer,W.**
R Heterojunctions in integrated optics
J. Vac. Sci. Technol. 14    186-194 (1977)

206 **Segmueller,A., Krishna,P., Esaki,L.**
X-ray diffraction study of a one-dimensional
GaAs-AlAs superlattice
J. Appl. Cryst. 10    1-6 (1977)

207 **Shah,J., Leheny,R.F., Wiegmann,W.**
Low-temperature absorption spectrum in GaAs in the
presence of optical pumping
Phys. Rev. B 16    1577-1580 (1977)

208 **Shmelev,G.M., Chaikovskii,I.A., Pavlovich,V.V., Epshtein,E.M.**
T Electron-phonon interaction in a superlattice
Phys. Status Solidi B 80    697-701 (1977)

209 **Tsang,W.T., Cho,A.Y.**
Growth of GaAs-Ga1-xAlxAs over preferentially etched
channels by molecular beam epitaxy: A technique
for two-dimensional thin-film definition
Appl. Phys. Lett. 30    293-296 (1977)

210  **Tsang,W.T., Ilegems,M.**
Selective area growth of GaAs/AlxGa1-xAs multilayer structures with molecular beam epitaxy using Si shadow masks
Appl. Phys. Lett. 31    301-304 (1977)

211  **Waho,T., Ogawa,S., Maruyama,S.**
GaAs(1-x)Sb(x) (0.3<x<0.9) grown by molecular beam epitaxy
Jpn. J. Appl. Phys. 16    1875-1876 (1977)

212  **Williamson,W.J.**
R  GaAs and related compounds for optical integrated circuits
Monitor 38    398-402 (1977)

213  **Yano,M., Nogami,M., Matsushima,Y., Kimata,M.**
Molecular beam epitaxial growth of InAs
Jpn. J. Appl. Phys. 16    2131-2137 (1977)

214  **Ziel,J.P. van der, Gossard,A.C.**
Absorption, refractive index, and birefringence of AlAs-GaAs monolayers
J. Appl. Phys. 48    3018-3023 (1977)

215  **Ziel,J.P. van der, Gossard,A.C.**
Absorption, refractive index, and birefringence of AlAs-GaAs monolayers
J. Appl. Phys. 48    3018-3023 (1977)

# Subject Categories and References
## Year 1978

**AlAs, GaAs, (AlGa)As**
216,217,218,219,220,221,222,223,224,225,226,227,228,229,
232,235,236,237,238,239,240,241,242,244,246,247,248,250,
255,256,257,259,260,261,262,263,268,269,271,272,275,276,
277,278,279,280,281,282,283,284,286,287,288

**GaSb, InAs, Ga(AsSb), (GaIn)As**
229,232,233,238,243,245,251,252,253,258,266,267,270,285

**InP, (AlIn)As, (GaIn)As, (GaIn)(PAs)**
229,230,231,232,233,238,243,246,252,253,266

**Growth Equipment**
219,226,227,229,230,231,232,234,238,247,249,254,255,256,
260,261,262,263,264,273,280

**Growth Mechanism, Surface Analysis**
223,229,230,231,232,233,234,238,243,244,245,246,247,249,
251,254,255,256,257,262,263,273,280,284,285

**Periodic Multilayer Structure**
217,218,228,229,235,258,259,260,261,265,266,267,268,269,
270,272,274,275,276,286,287,288

**Modulation Doping**
228,275

**Electrical Properties**
216,219,220,221,222,223,225,226,228,229,230,232,236,237,
238,239,241,242,246,247,248,250,251,252,254,255,260,261,
263,265,270,271,274,275,276,277,278,279,281,282,283,284,
285

**Optical Properties**
216,217,218,220,224,229,232,234,237,239,240,247,249,252,
253,254,255,261,263,266,268,269,272,276,277,285,286,287,
288

**Microwave Devices**
225,229,241,242,255,260,261,271,281,282

**Optoelectronic Devices**
224,237,248,252,253,260,272,276,277

216   **Abstreiter,G., Bauser,E., Fischer,A., Ploog,K.**
      Raman spectroscopy - a versatile tool for
          characterization of thin films and
          heterostructures of GaAs and Al(x)Ga(1-x)As
      Appl. Phys. 16    345-352 (1978)

217   **Andreoni,W., Baldereschi,A., Car,R.**
  T   Effects of cation order on the energy bands of
          GaAs-AlAs heterostructures
      Solid State Commun. 27    821-824 (1978)

218   **Barker,A.S.,Jr., Merz,J.L., Gossard,A.C.**
      Study of zone-folding effects on phonons in
          alternating monolayers of GaAs-AlAs
      Phys. Rev. B 17    3181-3196 (1978)

219   **Barnes,P.A., Cho,A.Y.**
      Nonalloyed Ohmic contacts to n-GaAs by molecular beam
          epitaxy
      Appl. Phys. Lett. 33    651-653 (1978)

220   **Calawa,A.R.**
      Effect of H2 on residual impurities in GaAs MBE layers
      Appl. Phys. Lett. 33    1020-1022 (1978)

221   **Casey,H.C.,Jr., Cho,A.Y., Lang,D.V.,
          Nicollian,E.H.**
      Measurement of MIS capacitors with oxygen-doped
          AlxGa1-xAs insulating layers on GaAs
      J. Vac. Sci. Technol. 15    1408-1411 (1978)

222   **Casey,H.C.,Jr., Cho,A.Y., Nicollian,E.H.**
      Use of oxygen-doped AlxGa1-xAs for the insulating
          layer in MIS structures
      Appl. Phys. Lett. 32    678-679 (1978)

223   **Cho,A.Y., Dernier,P.D.**
      Single-crystal-aluminium Schottky-barrier diodes
          prepared by molecular beam epitaxy (MBE) on GaAs
      J. Appl. Phys. 49    3328-3332 (1978)

224   **Conwell,B.M., Burnham,R.D.**
  R   Materials for integrated optics: GaAs
      Ann. Rev. Mater. Sci. 8    135-179 (1978)

225   **Covington,D.W., Hicklin,W.H.**
      p+-n hyperabrupt GaAs varactors grown by
          molecular-beam epitaxy
      Electron. Lett. 14    752-753 (1978)

226   **Covington,D.W., Meeks,E.L.**
      The growth of unintentionally doped layers of GaAs
          using molecular beam epitaxy
      Proc. Southeastcon '78 Reg. 3 Conf. (IEEE, New York,
          1978)    380-383 (1978)

227  Delhomme,B.J., Blanchet,R.C., Fumey,M., Urgell,J.J.
Emissivity determination of semiconductor substrates to measure their temperatures with an infrared thermometer system
Proc. Int. Symp. MECO 78, Athens, June 26 – 29, 1978, Eds. M. H. Hamza and S. G. Tzafestas (Acta Press, Anaheim, Zurich, 1978)  6-9 (1978)

228  Dingle,R., Stoermer,H.L., Gossard,A.C., Wiegmann,W.
Electron mobilities in modulation-doped semiconductor heterojunction superlattices
Appl. Phys. Lett. 33  665-667 (1978)

229  Esaki,L.
R  Semiconductor devices in perspective – discovery of and recent developments in tunneling devices
Phys. Contemp. Needs 2  29-87 (1978)

230  Farrow,R.F.C., Cullis,A.G., Grant,A.J., Pattison,J.E.
Structural and electrical properties of epitaxial metal films grown on argon ion bombarded and annealed (001)InP
J. Cryst. Growth 45  292-301 (1978)

231  Farrow,R.F.C., Williams,G.M.
A high temperature high purity source for metal beam epitaxy
Thin Solid Films 55  303-315 (1978)

232  Foxon,C.T.
R  Molecular beam epitaxy
Acta Electron. 21  139-150 (1978)

233  Foxon,C.T., Joyce,B.A.
Surface processes controlling the growth of $Ga_xIn_{1-x}As$ and $Ga_xIn_{1-x}P$ alloy films by MBE
J. Cryst. Growth 44  75-83 (1978)

234  Gonda,S., Matsushima,Y., Mukai,S., Makita,Y., Igarashi,O.
Heteroepitaxial growth of GaP on silicon by molecular beam epitaxy
Jpn. J. Appl. Phys. 17  1043-1048 (1978)

235  Herman,F., Kasowski,R.V.
T  Electronic structure of (110) Ge-GaAs superlattices and interfaces
Phys. Rev. B 17  672-674 (1978)

236  Hirose,M., Fischer,A., Ploog,K.
Growth of $Al_2O_3$ layer on MBE GaAs
Phys. Status Solidi A 45  K175-K177 (1978)

237 **Ilegems,M., Schwartz,B., Koszi,L.A., Miller,R.C.**
Integrated multijunction GaAs photodetector with high output voltage
Appl. Phys. Lett. 33   629-631 (1978)

238 **Joyce,B.A., Foxon,C.T., Neave,J.H.**
R Fundamentals of molecular beam epitaxy
J. Jpn. Assoc. Crystal Growth 5   185-197 (1978)

239 **Kroemer,H., Chien,W.Y., Casey,H.C.,Jr., Cho,A.Y.**
Photocollection efficiency and interface charges of MBE-grown abrupt p(GaAs)-N(Al0.33Ga0.67As)heterojunctions
Appl. Phys. Lett. 33   749-751 (1978)

240 **Lee,T.P., Holden,W.S., Cho,A.Y.**
AlGaAs-GaAs double-heterostructure small-area light-emitting diodes by molecular-beam epitaxy
Appl. Phys. Lett. 32   415-417 (1978)

241 **Linke,R.A., Schneider,M.V., Cho,A.Y.**
Cryogenic millimeter-wave receiver using molecular beam epitaxy diodes
IEEE Trans. Microwave Theory Techn. MTT-26   935-938 (1978)

242 **Linke,R.A., Schneider,M.V., Cho,A.Y.**
Cryogenic millimeter-wave receiver using molecular beam epitaxy diodes
1978 IEEE MTT-S Int. Microwave Symp. Digest   396-398 (1978)

243 **Ludeke,R.**
R Electronic properties of (100) surfaces of GaSb and InAs and their alloys with GaAs
IBM J. Res. Develop. 22   304-314 (1978)

244 **Ludeke,R., Ley,L., Ploog,K.**
Valence and core level photoemission spectra of Al(x)Ga(1-x)As
Solid State Commun. 28   57-60 (1978)

245 **Maruyama,S., Waho,T., Ogawa,S.**
Surface structure of GaAs(1-x)Sb(x) grown by molecular beam epitaxy
Jpn. J. Appl. Phys. 17   1695-1696 (1978)

246 **Massies,J., Devoldere,P., Linh,N.T.**
Silver contact on GaAs(001) and InP(001)
J. Vac. Sci. Technol. 15   1353-1357 (1978)

247 **Matsunaga,N., Suzuki,T., Takahashi,K.**
Ionized beam doping in molecular-beam epitaxy of GaAs and Al(x)Ga(1-x)As
J. Appl. Phys. 49   5710-5715 (1978)

248  **Matsunaga,N., Takahashi,K.**
Graded band-gap p-Al(x)Ga(1-x)As/n-GaAs heterojunction solar cells prepared by molecular beam epitaxy
Int. J. Electron. 45    273-282 (1978)

249  **Matsushima,Y., Gonda,S., Makita,Y., Mukai,S.**
Nitrogen doping into GaAs(1-x)P(x) using ionized beam in molecular beam epitaxy
J. Cryst. Growth 43    281-286 (1978)

250  **McLevige,W.V., Vaidyanathan,K.V., Streetman,B.G., Ilegems,M., Comas,J., Plew,L.**
Annealing studies of Be-doped GaAs grown by molecular beam epitaxy
Appl. Phys. Lett. 33    127-129 (1978)

251  **Meggitt,B.T., Parker,E.H.C., King,R.M.**
Thin InAs epitaxial layers grown on (100) GaAs substrates by molecular beam deposition
Appl. Phys. Lett. 33    528-530 (1978)

252  **Miller,B.I., McFee,J.H.**
Growth of $Ga_yIn_{1-y}As/InP$ heterostructures by molecular beam epitaxy
J. Electrochem. Soc. 125    1310-1317 (1978)

253  **Miller,B.I., McFee,J.H., Martin,R.J., Tien,P.K.**
Room-temperature operation of lattice-matched InP/Ga0.47In0.53As/InP double-heterostructure lasers grown by MBE
Appl. Phys. Lett. 33    44-47 (1978)

254  **Morimoto,K., Watanabe,H., Itoh,S.**
Ionized-cluster beam epitaxial growth of GaP films on GaP and Si substrates
J. Cryst. Growth 45    334-339 (1978)

255  **Murotani,T., Shimanoe,T., Mitsui,S.**
Growth temperature dependence in molecular beam epitaxy of gallium arsenide
J. Cryst. Growth 45    302-308 (1978)

256  **Neave,J.H., Joyce,B.A.**
Temperature range for growth of autoepitaxial GaAs by MBE
J. Cryst. Growth 43    204-208 (1978)

257  **Neave,J.H., Joyce,B.A.**
Structure and stoichiometry of (100) GaAs surfaces during molecular beam epitaxy
J. Cryst. Growth 44    387-397 (1978)

258 **Nucho,R.N., Madhukar,A.**
T Tight-binding study of the electronic structure of the InAs-GaSb(001) superlattice
J. Vac. Sci. Technol. 15 1530-1534 (1978)

259 **Petroff,P.M., Gossard,A.C., Wiegmann,W., Savage,A.**
Crystal growth kinetics in (GaAs)n-(AlAs)m superlattices deposited by molecular beam epitaxy. I. Growth on singular (100) GaAs substrates.
J. Cryst. Growth 44 5-13 (1978)

260 **Ploog,K.**
R Molekularstrahl-Epitaxie - Grundlagen und Anwendung fuer die Bauelementherstellung
Nachrichtentechn. Zeitschr. NTZ 31 435-441 (1978)

261 **Ploog,K., Doehler,G.H.**
R Dotierungsstrukturen in Galliumarsenidschichten - Herstellung und Verwendung in Bauelementen
Elektronik Industrie 9, Heft 3 11-13, Heft 4 13-15 (1978)

262 **Ploog,K., Fischer,A.**
R Molekularstrahl-Epitaxie von GaAs und in-situ Charakterisierung mit RHEED, SIMS und AES
Vakuum-Technik 27 162-171 (1978)

263 **Ploog,K., Fischer,A.**
Surface segregation of Sn during MBE of n-type GaAs established by SIMS and AES
J. Vac. Sci. Technol. 15 255-259 (1978)

264 **Robinson,J.W., Ilegems,M.**
Vacuum interlock system for molecular beam epitaxy
Rev. Sci. Instrum. 49 205-207 (1978)

265 **Romanov,Y.A., Bovin,V.P., Orlov,L.K.**
T Nonlinear amplification of electromagnetic oscillations in semiconductors with superlattices
Sov. Phys. Semicond. 12 987-989 * Fiz. Tekh. Poluprovodn. 12 1665-1669 (1978)

266 **Sai-Halasz,G.A., Chang,L.L., Welter,J.M., Chang,C.A., Esaki,L.**
Optical absorption of $In_{(1-x)}Ga_{(x)}-GaSb_{(1-y)}As_{(y)}$ superlattices
Solid State Commun. 27 935-937 (1978)

267 **Sai-Halasz,G.A., Esaki,L., Harrison,W.A.**
T InAs-GaSb superlattice energy structure and its semiconductor-semimetal transition
Phys. Rev. B 18 2812-2818 (1978)

268  Sai-Halasz,G.A., Pinczuk,A., Yu,P.Y., Esaki,L.
Superlattice umklapp processes in resonant Raman
    scattering
Surf. Sci. 73    232-237 (1978)

269  Sai-Halasz,G.A., Pinczuk,A., Yu,P.Y., Esaki,L.
Resonance enhanced umklapp Raman processes in
    GaAs-Ga(1-x)Al(x)As superlattices
Solid State Commun. 25    381-384 (1978)

270  Sakaki,H., Chang,L.L., Sai-Halasz,G.A., Chang,C.A.,
    Esaki,L.
Two-dimensional electronic structure in InAs-GaSb
    superlattices
Solid State Commun. 26    586-592 (1978)

271  Schneider,M.V.
Low-noise millimeter-wave Schottky mixers
Proc. 8th European Microwave Conf. (Microwave
    Exhibitions and Publishers Ltd, Sevenoaks,
    England, 1978)    682-688 (1978)

272  Shellan,J.B., Ng,W., Yeh,P., Yariv,A., Cho,A.Y.
Transverse Bragg-reflector injection lasers
Opt. Lett. 2    136-138 (1978)

273  Shen,L.Y.L.
Angular distribution of molecular beams from modified
    Knudsen cells for molecular beam epitaxy
J. Vac. Sci. Technol. 15    10-12 (1978)

274  Shmelev,G.M., Chaikovskii,I.A., Bau,N.K.
T    High-frequency conductivity of semiconductors with
    superlattice
Sov. Phys. Semicond. 12    1149-1152    *    Fiz. Tekh.
    Poluprovodn. 12    1932-1937 (1978)

275  Stoermer,H.L., Dingle,R., Gossard,A.C.,
    Wiegmann,W., Logan,R.A.
Electronic properties of modulation-doped
    GaAs-AlxGa1-xAs superlattices
Inst. Phys. Conf. Ser. 43    557-560 (1978)

276  Suris,R.A., Fedirko,V.A.
T    Heating photoconductivity in a semiconductor with a
    superlattice
Sov. Phys. Semicond. 12    629-632    *    Fiz. Tekh.
    Poluprovodn. 12    1060-1065 (1978)

277  Tsang,W.T.
The influence of bulk nonradiative recombination in
    the wide band-gap regions of molecular beam
    epitaxially grown GaAs-AlxGa1-xAs DH lasers
Appl. Phys. Lett. 33    245-248 (1978)

278   Tsang,W.T.
      Self-terminating thermal oxidation of AlAs epilayers
         grown on GaAs by molecular beam epitaxy
      Appl. Phys. Lett. 33   426-429 (1978)

279   Tsang,W.T.
      In situ Ohmic-contact formation to n-and p-GaAs by
         molecular beam epitaxy
      Appl. Phys. Lett. 33   1022-1025 (1978)

280   Tsang,W.T., Cho,A.Y.
      Molecular beam epitaxial writing of patterned GaAs
         epilayer structures
      Appl. Phys. Lett. 32   491-493 (1978)

281   Ury,I., Holm-Kennedy,J.W.
 T    Two-dimensional subbanding in junction field effect
         structures
      Surf. Sci. 73   179-189 (1978)

282   Wataze,M., Mitsui,Y., Shimanoe,T., Nakatani,M.,
         Mitsui,S.
      High-power GaAs F.E.T. prepared by molecular-beam
         epitaxy
      Electron. Lett. 14   759-761 (1978)

283   Wood,C.E.C.
      "Surface exchange" doping of MBE GaAs from S and Se
         "captive sources"
      Appl. Phys. Lett. 33   770-772 (1978)

284   Wood,C.E.C., Joyce,B.A.
      Tin-doping effects in GaAs films grown by molecular
         beam epitaxy
      J. Appl. Phys. 49   4854-4861 (1978)

285   Yano,M., Suzuki,Y., Ishii,T., Matsushima,Y.,
         Kimata,M.
      Molecular beam epitaxy of GaSb and GaSb(x)As(1-x)
      Jpn. J. Appl. Phys. 17   2091-2096 (1978)

286   Yeh,P., Yariv,A., Cho,A.Y.
      Optical surface waves in periodic layered media
      Appl. Phys. Lett. 32   104-105 (1978)

287   Ziel,J.P. van der, Gossard,A.C.
      Optical birefringence of ultrathin
         Al(x)Ga(1-x)As-GaAs multilayer heterostructures
      J. Appl. Phys. 49   2919-2921 (1978)

288   Ziel,J.P. van der, Gossard,A.C.
      Two-photon absorption spectrum of AlAs-GaAs monolayer
         crystals
      Phys. Rev. B 17   765-768 (1978)

# Subject Categories and References
# Year 1979

### AlAs, GaAs, (AlGa)As
289,290,291,292,294,295,296,297,298,299,301,302,303,304,
305,307,309,310,311,312,314,315,316,317,318,320,321,322,
323,324,325,326,327,328,331,332,333,334,337,338,339,342,
343,344,345,348,349,350,351,352,353,355,357,358,359,360,
361,362,363,364,365,366,367,368,369,370,371,372,373,374,
375,376,377,378,379,380,383,384

### GaSb, InAs, Ga(AsSb), (GaIn)As
293,294,300,306,307,308,318,319,330,335,336,339,341,346,
347,364

### InP, (AlIn)As, (GaIn)As, (GaIn)(PAs)
293,306,307,309,313,321,335,340,341,356,364

### Growth Equipment
293,299,309,317,322,331,343,345,350,351,356,360,367,368,
373,385

### Growth Mechanism, Surface Analysis
293,294,295,299,309,313,317,321,338,339,342,343,344,345,
348,349,350,351,352,353,356,360,362,368,375,379,380,381,
382,385

### Periodic Multilayer Structure
290,291,292,300,301,307,308,309,315,316,317,318,319,328,
332,333,336,337,346,347,352,354,355,357,358,359,362,364,
365,366,377

### Modulation Doping
289,290,309,315,332,333,358,359,361,371,372

### Electrical Properties
289,290,292,293,294,295,296,297,298,303,304,305,307,308,
309,310,311,312,314,315,316,317,318,320,321,322,323,330,
331,332,333,334,336,337,339,340,344,346,347,350,353,354,
356,360,362,363,364,365,366,367,368,369,370,371,372,373,
374,376,378,379,380,381,383,384

### Optical Properties
289,290,291,292,293,297,300,301,302,307,311,312,315,316,
317,324,325,326,327,328,331,335,337,340,347,352,355,356,
358,359,360,363,364,367,368,372,373,374,377,380,382

### Microwave Devices
296,298,305,309,314,319,322,334,361,380

### Optoelectronic Devices
301,309,319,320,331,335,373,374,377

289 **Abstreiter,G., Ploog,K.**
Inelastic light scattering from a quasi-two-dimensional electron system in GaAs-Al(x)Ga(1-x)As heterojunctions
Phys. Rev. Lett. 42   1308-1311 (1979)

290 **Ando,T., Mori,S.**
T  Electronic properties of a semiconductor superlattice.I.Self-consistent calculation of subband structure and optical spectra
J. Phys. Soc. Jpn. 47   1518-1528 (1979)

291 **Andreoni,W., Car,R., Baldereschi,A.**
T  Effects of cation order on the energy bands of GaAs-AlAs heterostructures
Inst. Phys. Conf. Ser. 43   733-736 (1979)

292 **Arthur,J.R.,Jr., Johnson,W.C.**
R  Superlattices
Ind. Res. Dev. 21 No. 1   109-116 (1979)

293 **Asahi,H., Okamoto,H., Ikeda,M., Kawamura,Y.**
Properties of molecular beam epitaxial In(x)Ga(1-x)As (x=0.53) layers grown on InP substrates
Jpn. J. Appl. Phys. 18   565-573 (1979)

294 **Bachrach,R.Z.**
R  Semiconductor surface and crystal physics studied by MBE
Prog. Crystal Growth Charact. 2   115-144 (1979)

295 **Bachrach,R.Z., Bauer,R.S., McMenamin,J.C., Bianconi,A.**
Microscopic aspects of metal-GaAs interface formation
Inst. Phys. Conf. Ser. 43   1073-1076 (1979)

296 **Bandy,S.G., Collins,D.M., Mishimoto,C.K.**
Low-noise microwave F.E.T.S. fabricated by molecular-beam epitaxy
Electron. Lett. 15  218-219 (1979)

297 **Bean,J.C., Dingle,R.**
Luminescent p-GaAs grown by zinc ion doped MBE
Appl. Phys. Lett. 35   925-927 (1979)

298 **Berson,B.**
R  Making 1980's semiconductors: molecules, electrons, and ions
MSN Microwave Syst. News 9 No. 9   45-56 (1979)

299 **Blanchet,R.C., Delhomme,B.J., Urgell,J.J.**
Analyse structurale du procede d'epitaxie par jets moleculaires: Application a l'arseniure de gallium
L'onde electrique 59   83-92 (1979)

300 Bluyssen,H.J.A., Maan,J.C., Wyder,P., Chang,L.L., Esaki,L.
Cyclotron resonance in an InAs-GaSb superlattice
Solid State Commun. 31    35-38 (1979)

301 Burnham,R.D., Scifres,D.R.
R Integrated optical devices fabricated by MBE
Prog. Crystal Growth Charact. 2    95-113 (1979)

302 Carenco,A., Menigaux,L., Alexandre,F., Abdalla,M.I., Brenac,A.
Directional coupler switch in molecular-beam epitaxy GaAs
Appl. Phys. Lett. 34    755-757 (1979)

303 Casey,H.C.,Jr., Cho,A.Y., Foy,P.W.
Reduction of surface recombination current with oxygen-doped Al0.5Ga0.5As surface layer on n-type GaAs
J. Vac. Sci. Technol. 16    1398-1401 (1979)

304 Casey,H.C.,Jr., Cho,A.Y., Foy,P.W.
Reduction of surface recombination current in GaAs p-n junctions
Appl. Phys. Lett. 34    594-596 (1979)

305 Casey,H.C.,Jr., Cho,A.Y., Lang,D.V., Nicollian,E.H., Foy,P.W.
Investigation of heterojunctions for MIS devices with oxygen-doped $Al_xGa_{1-x}As$ on n-type GaAs
J. Appl. Phys. 50    3484-3491 (1979)

306 Chang,C.A., Segmueller,A.
Substrate effect on the lattice constants of the MBE-grown $In_{(1-x)}Ga_{(x)}As$ and $GaSb_{(1-y)}As_{(y)}$
J. Vac. Sci. Technol. 16    285-286 (1979)

307 Chang,L.L., Esaki,L.
R Semiconductor superlattices by MBE and their characterization
Prog. Crystal Growth Charact. 2    3-14 (1979)

308 Chang,L.L., Kawai,N.J., Sai-Halasz,G.A., Ludeke,R., Esaki,L.
Observation of semiconductor-semimetal transition in InAs-GaSb superlattices
Appl. Phys. Lett. 35    939-942 (1979)

309 Cho,A.Y.
R Recent developments in molecular beam epitaxy (MBE)
J. Vac. Sci. Technol. 16    275-284 (1979)

310 Collins,D.M.
The use of SnTe as the source of donor impurities in GaAs grown by molecular beam epitaxy
Appl. Phys. Lett. 35    67-70 (1979)

**311** Covington,D.W., Litton,C.W., Reynolds,D.C., Almassy,R.J., McCoy,G.L.
R  Photoluminescence and electrical characterization of MBE GaAs epilayers: a recent survey
Inst. Phys. Conf. Ser. 45  171-180 (1979)

**312** Covington,D.W., Meeks,E.L.
R  Unintentional dopants incorporated in GaAs layers grown by molecular beam epitaxy
J. Vac. Sci. Technol. 16  847-850 (1979)

**313** Cullis,A.G., Farrow,R.F.C.
A study of the structure and properties of epitaxial silver deposited by atomic beam techniques on (001)InP
Thin Solid Films 58  197-202 (1979)

**314** DiLorenzo,J.V., Niehaus,W.C., Cho,A.Y.
Nonalloyed and in situ Ohmic contacts to highly doped n-type GaAs layers grown by molecular beam epitaxy (MBE) for field-effect transistors
J. Appl. Phys. 50  951-954 (1979)

**315** Dingle,R., Stoermer,H.L., Gossard,A.C., Wiegmann,W.
R  Electronic properties of modulation-doped GaAs-AlGaAs heterojunction superlattices
Inst. Phys. Conf. Ser. 45  248-255 (1979)

**316** Doehler,G.H.
T  Doping superlattices
J. Vac. Sci. Technol. 16  851-856 (1979)

**317** Doehler,G.H., Ploog,K.
R  Periodic doping structure in GaAs
Progr. Crystal Growth Charact. 2  145-168 (1979)

**318** Esaki,L., Chang,L.L.
R  Semiconductor superlattices in high magnetic fields
J. Magn. Magnetic Mater. 11  208-215 (1979)

**319** Esaki,L., Sai-Halasz,G.A.
T  Novel microwave/infrared detectors with semiconductor superlattices
IBM Tech. Disclosure Bull. 22  1262-1264 (1979)

**320** Fan,J.C.C., Calawa,A.R., Chapman,R.L., Turner,G.W.
Efficient shallow-homojunction GaAs solar cells by molecular beam epitaxy
Appl. Phys. Lett. 35  804-806 (1979)

**321** Farrow,R.F.C., Cullis,A.G., Grant,A.J., Jones,G.R.
Molecular beam epitaxy and field emission deposition for metal film growth on III-V compound semiconductors - a comparative study
Thin Solid Films 58  189-196 (1979)

322  Fujii,T., Suzuki,H., Hiyamizu,S.
     Sn-doped GaAs films grown by molecular beam epitaxy
     Fujitsu Sci. Tech. J.    121-130 (1979)

323  Garner,C.M., Su,C.Y., Shen,Y.D., Lee,C.S.,
         Pearson,G.L., Spicer,W.E.
     Interface studies of Al(x)Ga(1-x)As-GaAs
         heterojunctions
     J. Appl. Phys. 50    3383-3389 (1979)

324  Gibbs,H.M., Gossard,A.C., McCall,S.L., Passner,A.,
         Wiegmann,W., Venkatesan,T.N.C.
     Saturation of the free exciton resonance in GaAs
     Solid State Commun. 30    271-275 (1979)

325  Gibbs,H.M., McCall,S.L., Gossard,A.C., Passner,A.,
         Wiegmann,W., Venkatesan,T.N.C.
     Controlling light with light: optical bistability and
         optical modulation
     in "Laser Spectroscopy IV", Eds. H. Walther and K.W.
         Rothe ( Springer Verlag, Berlin, Heidelberg, New
         York, 1979 )    441-450 (1979)

326  Gibbs,H.M., McCall,S.L., Venkatesan,T.N.C.,
         Gossard,A.C., Passner,A., Wiegmann,W.
     Optical bistability in semiconductors
     Appl. Phys. Lett. 35    451-453 (1979)

327  Gibbs,H.M., Venkatesan,T.N.C., McCall,S.L.,
         Passner,A., Gossard,A.C., Wiegmann,W.
     Optical modulation by optical tuning of a cavity
     Appl. Phys. Lett. 34    511-514 (1979)

328  Gossard,A.C.
  R  GaAs/AlAs layered films
     Thin Solid Films 57    3-13 (1979)

329  Grange,J.D., Parker,E.H.C.
  R  Device fabrication for the future?
     Phys. Bull. 30    20-22 (1979)

330  Grange,J.D., Parker,E.H.C., King,R.M.
     Relationship of MBE growth parameters with the
         electrical properties of thin (100) InAs epilayers
     J. Phys. D 12    1601-1612 (1979)

331  Gulyaev,Y.V., Dvoryankina,G.G., Dvoryankin,V.F.,
         Cherevatskii,N.Y.
  R  Molecular beam epitaxy - a promising method for
         fabrication of integrated-optics devices. I.
         Injection lasers
     Sov. J. Quantum Electron. 9    1-12    *    Kvantovaya
         Electron. (Moskow) 6    5-24 (1979)

332 **Hess,K.**
T   Impurity and phonon scattering in layered structures
    Appl. Phys. Lett. 35   484-486 (1979)

333 **Hess,K., Morkoc,H., Shichijo,H., Streetman,B.G.**
    Negative differential resistance through real-space electron transfer
    Appl. Phys. Lett. 35   469-471 (1979)

334 **Hierl,T.L., Collins,D.M.**
    GaAs Read IMPATT diodes by molecular beam epitaxy
    Proc.-Bienn. Cornell Electr. Eng. Conference 7th 369-377 (1979)

335 **Hiyamizu,S., Fujii,T., Nanbu,K., Maekawa,S., Hisatsugu,T.**
    Laser operation of heteroepitaxial $In_xGa_{1-x}As$ by molecular beam epitaxy
    Surf. Sci. 86   137-143 (1979)

336 **Ihm,J., Lam,P.K., Cohen,M.L.**
T   Electronic structure of the (001)InAs-GaSb superlattice
    Phys. Rev. B 20   4120-4125 (1979)

337 **Ivanov,I., Pollmann,J.**
T   Microscopic approach to the quantum size effect in superlattices
    Solid State Commun. 32   869-872 (1979)

338 **Jacobi,K., Muschwitz,C. v., Ranke,W.**
    Angular resolved UPS of surface states on GaAs(111) prepared by molecular beam epitaxy
    Surf. Sci. 82   270-282 (1979)

339 **Joyce,B.A.**
R   Present status and future directions for MBE
    Surf. Sci. 86   92-101 (1979)

340 **Kawamura,Y., Ikeda,M., Asahi,H., Okamoto,H.**
    Photoluminescence of undoped (100)InP homoepitaxial films grown by molecular beam epitaxy
    Appl. Phys. Lett. 35   481-484 (1979)

341 **Kawamura,Y., Okamoto,H.**
    Lattice deformation and misorientation of $In(x)Ga(1-x)As$ epitaxial layers grown on InP substrates by molecular beam epitaxy
    J. Appl. Phys. 50   4457-4458 (1979)

342 **Larsen,P.K., Neave,J.H., Joyce,B.A.**
    Angular resolved photoemission from surface states on reconstructed (100) GaAs surfaces
    J. Phys. C 12   L869-L874 (1979)

343  **Laurence,G., Simondet,F., Saget,P.**
Combined RHEED-AES study of the thermal treatment of
(001) GaAs surface prior to MBE growth
Appl. Phys. Lett. 19   63-70 (1979)

344  **Ludeke,R., Ley,L.**
Surface effects in X-ray photoemission from GaAs
Inst. Phys. Conf. Ser. 43   1069-1072 (1979)

345  **Luscher,P.E., Collins,D.M.**
R   Design considerations for molecular beam epitaxy
systems
Prog. Crystal Growth Charact. 2   15-32 (1979)

346  **Madhukar,A., Dandekar,N.V., Nucho,R.N.**
T   Two-dimensional effects and effective masses of the
InAs/GaSb(001) superlattices
J. Vac. Sci. Technol. 16   1507-1511 (1979)

347  **Madhukar,A., Nucho,R.N.**
T   The electronic structure of InAs/GaAs(001)
superlattices - two dimensional effects
Solid State Commun. 32   331-336 (1979)

348  **Marra,W.C., Eisenberger,P., Cho,A.Y.**
X-ray total-external-reflection-Bragg diffraction: A
structural study of the GaAs-Al interface
J. Appl. Phys. 50   6927-6933 (1979)

349  **Massies,J., Chaplart,J., Linh,N.T.**
New results in the study of the aluminium epitaxial
growth on gallium arsenide (001)
Solid State Commun. 32   707-709 (1979)

350  **Massies,J., Devoldere,P., Linh,N.T.**
Work function measurements on MBE GaAs(001) layers
J. Vac. Sci. Technol. 16   1244-1247 (1979)

351  **Massies,J., Etienne,P., Linh,N.T.**
The interaction of silver and aluminium on gallium
arsenide (001) surface: a study by MBE and
associated techniques
Surf. Sci. 80   550-556 (1979)

352  **Miller,R.C., Kleinman,D.A., Gossard,A.C.**
Electron spin orientation in optically pumped
$GaAs-Al_xGa_{1-x}As$ multilayer structures
Inst. Phys. Conf. Ser. 43   1043-1046 (1979)

353  **Morkoc,H., Cho,A.Y.**
High-purity GaAs and Cr-doped GaAs epitaxial layers
by MBE
J. Appl. Phys. 50   6413-6416 (1979)

354 **Nakao,K.**
T Electronic band structure of superlattices in magnetic fields
J. Phys. Soc. Jpn. 46 1669-1670 (1979)

355 **Narayanamurti,V., Stoermer,H.L., Chin,M.A., Gossard,A.C., Wiegmann,W.**
Selective transmission of high-frequency phonons by a superlattice: the "dielectric" phonon filter
Phys. Rev. Lett. 43 2012-2016 (1979)

356 **Norris,M.T., Stanley,C.R.**
Substrate temperature limits for epitaxy of InP by MBE
Appl. Phys. Lett. 35 617-620 (1979)

357 **Petroff,P.M., Gossard,A.C., Savage,A., Wiegmann,W.**
Molecular beam epitaxy of Ge and Ga1-xAlxAs ultra thin film superlattices
J. Cryst. Growth 46 172-178 (1979)

358 **Pinczuk,A., Stoermer,H.L., Dingle,R., Worlock,J.M., Wiegmann,W., Gossard,A.C.**
Observation of intersubband excitations in a multilayer two dimensional electron gas
Solid State Commun. 32 1001-1003 (1979)

359 **Pinczuk,A., Stoermer,H.L., Dingle,R., Worlock,J.M., Wiegmann,W., Gossard,A.C.**
R Inelastic light scattering by the two dimensional electrons in semiconductor heterojunction
in "Light Scattering in Solids", Eds. J.L.Birman, H.Z.Cummins, K.K.Rebance (Plenum Press, New York, 1979) 307-314 (1979)

360 **Ploog,K.**
R Surface studies during molecular beam epitaxy of gallium arsenide
J. Vac. Sci. Technol. 16 838-846 (1979)

361 **Ploog,K.**
R Anreicherung von Ladungstraegern an der Grenzflaeche zweier Halbleiter mit unterschiedlichem Bandabstand
Elektronik Industrie 10, Heft 11 13-14 (1979)

362 **Ploog,K., Fischer,A., Kuenzel,H.**
Improved p/n junctions in Ge-doped GaAs grown by molecular beam epitaxy
Appl. Phys. 18 353-356 (1979)

363 **Ploog,K., Fischer,A., Trommer,R., Hirose,M.**
MBE-grown insulating oxide films on GaAs
J. Vac. Sci. Technol. 16 290-294 (1979)

364 **Sai-Halasz,G.A.**
R Semiconductor superlattices
Inst. Phys. Conf. Ser. 43 21-30 (1979)

365 Sakaki,H., Chang,L.L., Esaki,L.
Subband-structure related anisotropy in negative
 magnetoresistivity of semiconductor superlattices
Inst. Phys. Conf. Ser. 43    737-740 (1979)

366 Schulman,J.N., McGill,T.C.
T  Electronic properties of the AlAs-GaAs(001) interface
 and superlattice
Phys. Rev. B 19    6341-6349 (1979)

367 Scott,G.B., Roberts,J.S.
Photoluminescence in III-V compounds grown by MBE
Inst. Phys. Conf. Ser. 45    181-189 (1979)

368 Shimanoe,T., Murotani,T., Nakatani,M., Otsubo,M.,
 Mitsui,S.
High quality Si-doped GaAs layers grown by molecular
 beam epitaxy
Surf. Sci. 86    126-136 (1979)

369 Spencer,M., Stall,R.A., Eastman,L.F., Wood,C.E.C.
Characterization of grain bounderies using deep level
 transient spectroscopy
J. Appl. Phys. 50    8006-8009 (1979)

370 Stall,R.A., Wood,C.E.C., Board,K., Eastman,L.F.
Ultra low resistance Ohmic contacts to n-GaAs
Electron. Lett. 15    800-801 (1979)

371 Stoermer,H.L., Dingle,R., Gossard,A.C.,
 Wiegmann,W., Sturge,M.D.
Two-dimensional electron gas at differentially doped
 GaAs-Al$_x$Ga$_{1-x}$As heterojunction interface
J. Vac. Sci. Technol. 16    1517-1519 (1979)

372 Stoermer,H.L., Dingle,R., Gossard,A.C.,
 Wiegmann,W., Sturge,M.D.
Two-dimensional electron gas at a
 semiconductor-semiconductor interface
Solid State Commun. 29    705-707 (1979)

373 Suzuki,T., Konagai,M., Takahashi,K.
Electrical properties of Zn+ doped (AlGa)As prepared
 by molecular beam epitaxy and its use in shallow
 junction solar cells
Thin Solid Films 60    85-89 (1979)

374 Tsang,W.T.
Low-current-threshold and high-lasing uniformity
 GaAs-Al$_x$Ga$_{1-x}$As double-heterostructure lasers
 grown by molecular beam epitaxy
Appl. Phys. Lett. 34    473-475 (1979)

375 **Tsang,W.T., Ilegems,M.**
The preparation of GaAs thin-film optical components by molecular beam epitaxy using Si shadow masking technique
Appl. Phys. Lett. 35   792-795 (1979)

376 **Tsang,W.T., Olmstead,M., Chang,R.P.H.**
Multidielectrics for GaAs MIS devices using composition-graded $Al_xGa_{1-x}As$ and oxidized AlAs
Appl. Phys. Lett. 34   408-410 (1979)

377 **Tsang,W.T., Weisbuch,C., Miller,R.C., Dingle,R.**
Current injection GaAs-$Al(x)Ga(1-x)As$ multi-quantum-well heterostructure lasers prepared by molecular beam epitaxy
Appl. Phys. Lett. 35   673-675 (1979)

378 **Weimann,G.**
Electrical characterization of n-type GaAs films grown by molecular beam epitaxy
Phys. Status Solidi A 53   K173-K176 (1979)

379 **Weimann,G., Schlapp,W.**
Molekularstrahl-Epitaxie von GaAs-Schichten
Technischer Bericht (FI beim FTZ) No. 65 TBr 17   1-38 (1979)

380 **Wood,C.E.C., Woodcock,J.M., Harris,J.J.**
Low-compensation n-type and flat-surface p-type Ge-doped GaAs by molecular beam epitaxy
Inst. Phys. Conf. Ser. 45   28-37 (1979)

381 **Yano,M., Takase,T., Kimata,M.**
Heteroepitaxial InSb films grown by molecular beam epitaxy
Phys. Status Solidi A 54   707-713 (1979)

382 **Yano,M., Takase,T., Kimata,M.**
Molecular beam epitaxy of $In(x)Ga(1-x)Sb$ ($0<=x<=1$)
Jpn. J. Appl. Phys. 18   387-388 (1979)

383 **Yokoyama,S., Yukitomo,K., Hirose,M., Osaka,Y.**
GaAs MOS structures with $Al_2O_3$ grown by molecular beam reaction under UV excitation
Thin Solid Films 56   81-88 (1979)

384 **Yokoyama,S., Yukitomo,K., Hirose,M., Osaka,Y., Fischer,A., Ploog,K.**
GaAs MOS structures with $Al_2O_3$ grown by molecular beam reaction
Surf. Sci. 86   835-840 (1979)

385 **Yoshida,S., Misawa,S., Fujii,Y., Takada,S., Hayakawa,H., Gonda,S., Itoh,A.**
Reactive molecular beam epitaxy of aluminium nitride
J. Vac. Sci. Technol. 16   990-993 (1979)

# Subject Categories and References
# Year 1980

**AlAs, GaAs, (AlGa)As**
386,387,388,389,392,393,395,396,399,401,402,403,404,405,
406,407,409,410,411,412,414,415,416,419,421,422,424,425,
426,427,428,429,430,431,432,434,435,436,437,438,439,440,
443,444,445,446,447,448,449,452,453,454,455,456,457,458,
459,460,461,462,463,464,465,471,472,473,474,475,476,477,
478,479,480,482,483,485,486,487,488,489,490,491,492,493,
494,495,496,497,498,499,500,501,502,504,505,506,507,508

**GaSb, InAs, Ga(AsSb), (GaIn)As**
390,397,398,399,400,408,413,417,423,433,434,441,442,450,
451,467,469,470,482,484,503,505

**InP, (AlIn)As, (GaIn)As, (GaIn)(PAs)**
390,397,399,400,408,434,441,442,451,466,467,469,470,473,
505

**Growth Equipment**
388,395,399,400,403,417,434,443,473,474,480,504,505

**Growth Mechanism, Surface Analysis**
387,388,393,399,400,402,406,408,414,415,420,430,437,438,
439,442,446,447,448,450,459,466,467,468,472,473,480,505

**Periodic Multilayer Structure**
391,398,399,400,401,405,407,411,412,417,419,421,422,423,
432,433,434,451,452,453,457,460,461,471,472,474,477,478,
479,480,481,482,484,491,496,501,502,503,505

**Modulation Doping**
386,407,409,411,412,422,429,434,458,460,461,464,477,478,
479,480,491,492,501,502,504,505

**Electrical Properties**
386,388,389,390,391,392,393,396,398,399,400,406,407,409,
410,411,412,413,416,417,418,421,422,423,424,425,427,428,
429,430,431,433,434,438,441,442,443,444,445,446,447,449,
450,454,455,456,458,461,462,463,464,465,466,469,470,471,
474,475,480,482,485,487,488,489,490,491,492,499,501,502,
504,505,506,507

**Optical Properties**
    386,391,396,398,400,401,403,405,406,407,411,417,422,423,
    434,435,438,439,440,441,449,451,452,453,456,457,472,474,
    475,476,477,478,479,480,482,485,486,489,491,493,494,495,
    496,497,498,499,500,503

**Microwave Devices**
    390,409,410,411,424,425,426,430,434,443,454,458,469,470,
    474,475,480,505,506

**Optoelectronic Devices**
    403,411,434,435,440,443,456,470,472,474,475,476,480,485,
    486,493,494,495,496,497,498,499,500,505,508

386 **Abstreiter,G.**
R  Electronic properties of the two-dimensional system
      at GaAs/Al$_x$Ga$_{1-x}$As interfaces
   Surf. Sci. 98    117-125 (1980)

387 **Aleksandrov,L.N.**
TR Impurity distribution in epitaxial films condensed
      from vapour or molecular ion beams
   J. Cryst. Growth 48    635-643 (1980)

388 **Alexandre,F., Raisin,C., Abdalla,M.I., Brenac,A.,
   Masson,J.M.**
   Influence of growth conditions on tin incorporation
      in GaAs grown by molecular beam epitaxy
   J. Appl. Phys. 51    4296-4304 (1980)

389 **Allyn,C.L., Gossard,A.C., Wiegmann,W.**
   New rectifying semiconductor structure by molecular
      beam epitaxy
   Appl. Phys. Lett. 36    373-376 (1980)

390 **Barnard,J.A., Ohno,H., Wood,C.E.C., Eastman,L.F.**
   Double heterostructure Ga0.47In0.53As MESFETs with
      submicron gates
   IEEE Electron Device Lett.  EDL-1    174-176 (1980)

391 **Bass,F.G., Lykakh,V.A., Tetervov,A.P.**
T  Cyclotron resonance in a superlattice semiconductor
   Sov. Phys. Semicond. 14    1372-1376    *    Fiz. Tekh.
      Poluprovodn. 14    2314-2322 (1980)

392 **Basu,P.K., Nag,B.R.**
T  Lattice scattering mobility of a two-dimensional
      electron gas in GaAs
   Phys. Rev. B 22    4849-4852 (1980)

393 **Blanchet,R.C., Delhomme,B.J., Urgell,J.J.**
   Deposition of SiO as insulating film on GaAs in a
      molecular beam epitaxy (MBE) system
   Vide 201    Suppl.    594-597 (1980)

394 **Blood,P., Bye,K.L., Roberts,J.S.**
   Composition studies of MBE GaInP alloys by Rutherford
      scattering and x-ray diffraction
   J. Appl. Phys. 51    1790-1797 (1980)

395 **Budarnykh,V.I., Logvinskiy,L.M., Nesterikhin,Y.E.,
   Ostapovskiy,L.M., Ryabchenko,V.E.,
   Tsukerman,V.G.**
   The development of a technological version of an
      automated molecular-beam epitaxy system
   Automatic Monitoring No.6    3-6    *    Avtometriya
      No.6    4-11 (1980)

396 **Chai,Y.G.**
Effect of accelerated growth rate (1-5 mue m/h) on molecular beam epitaxial GaAs using Si as a dopant
Appl. Phys. Lett. 37   379-382 (1980)

397 **Chang,C.A., Serrano,C.M., Chang,L.L., Esaki,L.**
Studies by cross-sectional transmission electron microscope of InAs grown by molecular beam epitaxy on GaAs substrates
Appl. Phys. Lett. 37   538-540 (1980)

398 **Chang,L.L.**
R Semiconductor-semimetal transitions in InAs-GaSb superlattices
J. Phys. Soc. Jpn. 49, Suppl. A   997-1004 (1980)

399 **Chang,L.L.**
R Molecular beam epitaxy
in "Handbook of Semiconductors", Ed. S.P. Keller (North-Holland, Amsterdam, 1980) Vol. 3   563-597 (1980)

400 **Chang,L.L., Esaki,L.**
R Electronic properties of InAs-GaSb superlattices
Surf. Sci. 98   70-89 (1980)

401 **Chaplik,A.V., Krasheninnikov,M.V.**
T Two-dimensional plasmons in multi-layered structures
Solid State Commun. 35   189-190 (1980)

402 **Chinen,K., Niigaki,M., Miyao,M., Hagino,M.**
GaAs transmission photocathode grown by MBE
Jpn. J. Appl. Phys. 19   L703-L706 (1980)

403 **Cho,A.Y., Casey,H.C.,Jr., Radice,C., Foy,P.W.**
Influence of growth conditions on the threshold current density of double-heterostructure lasers prepared by molecular beam epitaxy
Electron. Lett. 16   72-74 (1980)

404 **Cho,A.Y., Tsang,W.T.**
Masked molecular beam epitaxy
IEEE, Optical Soc. America, Top. Meet. Integrated and Guided-Wave Optics, Technical Digest Wb1   1-6 (1980)

405 **Colvard,C., Merlin,R., Klein,M.V., Gossard,A.C.**
Observation of folded acoustic phonons in a semiconductor superlattice
Phys. Rev. Lett. 45   298-301 (1980)

406 **Covington,D.W., Comas,J., Yu,P.W.**
Iron doping in gallium arsendide by molecular beam epitaxy
Appl. Phys. Lett. 37   1094-1096 (1980)

407 Das Sarma,S., Madhukar,A.
T   Study of electron-phonon interaction and
       magneto-optical anomalies in two-dimensionally
       confined systems
    Phys. Rev. B 22    2823-2836 (1980)

408 Davies,G.J., Heckingbottom,R., Ohno,H.,
       Wood,C.E.C., Calawa,A.R.
    Arsenic stabilization of InP substrates for growth of
       Ga(x)In(1-x)As layers by molecular beam epitaxy
    Appl. Phys. Lett. 37    290-292 (1980)

409 Delagebeaudeuf,D., Delescluse,P., Etienne,P.,
       Laviron,M., Chaplart,J., Linh,N.T.
    Two-dimensional electron gas M.E.S.F.E.T. structure
    Electron. Lett. 16    667-668 (1980)

410 Devlin,W.J., Wood,C.E.C., Stall,R.A., Eastman,L.F.
    A molybdenium source, gate and drain metallization
       system for GaAs MESFET layers grown by molecular
       beam epitaxy
    Solid-State Electron. 23    823-829 (1980)

411 Dingle,R., Gossard,A.C., Stoermer,H.L.
 R  Building semiconductors from the atom up
    Bell Labs. Record 58    274-281 (1980)

412 Dingle,R., Stoermer,H.L., Gossard,A.C.,
       Wiegmann,W.
 R  Electronic properties of the GaAs-AlGaAs interface
       with applications to multi-interface
       heterojunction superlattices
    Surf. Sci. 98    90-100 (1980)

413 Doehler,G.H.
 T  Electron-hole subbands at the GaSb-InAs interface
    Surf. Sci. 98    108-116 (1980)

414 Dorfman,V.F.
 T  A microscopic mechanism of film growth from
       noncondensed phases
    Thin Solid Films 66    91-110 (1980)

415 Dvoryankin,V.F., Dvoryankina,G.G.,
       Cherevatskii,N.Y.
    Structure and chemical composition of surfaces of
       substrates and autoepitaxial films of GaAs (100)
       obtained by means of molecular beam epitaxy
    Inorg. Mater. 16    1432-1434    *    Izvest. Akad. Nauk
       SSSR, Neorg. Mater. 16    2103-2106 (1980)

416 Eastman,L.F., Stall,R.A., Woodard,D.W.,
       Dandekar,N., Wood,C.E.C., Shur,M.S., Board,K.
    Ballistic electron motion in GaAs at room temperature
    Electron. Lett. 16    524-525 (1980)

417 Esaki,L.
R   InAs-GaSb superlattices - synthesized narrow-gap
    semiconductors and semimetals
    Lect. Notes Phys. 133   302-323 (1980)

418 Farrow,R.F.C., Robertson,D.S., Williams,G.M.,
    Cullis,A.G., Jones,G.R., Young,I.M.
    The preparation of metastable, heteroepitaxial films
    of alpha-Sn by metal beam epitaxy
    Proc. 8th Int. Vac. Congr., Eds. F.Abeles and
    M.Croset (Soc. Franc. Vide, Paris 1980) Vol. 1
    109-112 (1980)

419 Fleming,R.M., McWhan,D.B., Gossard,A.C.,
    Wiegmann,W., Logan,R.A.
    X-ray diffraction study of interdiffusion and growth
    in $(GaAs)_n(AlAs)_m$ multilayers
    J. Appl. Phys. 51   357-363 (1980)

420 Foxon,C.T., Joyce,B.A., Norris,M.T.
    Composition effects in the growth of $Ga(In)As_yP_{1-y}$
    alloys by MBE
    J. Cryst. Growth 49   132-140 (1980)

421 Glisson,T.H., Hauser,J.R., Littlejohn,M.A.,
    Hess,K., Streetman,B.G., Shichijo,H.
T   Monte Carlo simulation of real-space electron
    transfer in GaAs-AlGaAs heterostructures
    J. Appl. Phys. 51   5445-5449 (1980)

422 Gornik,E., Schawarz,R., Tsui,D.C., Gossard,A.C.,
    Wiegmann,W.
    Far infrared emission from 2D electrons at the
    $GaAS-Al_xGa_{1-x}As$ interface
    J. Phys. Soc. Jpn. 49 Suppl.A   1029-1032 (1980)

423 Guldner,Y., Vieren,J.P., Voisin,P., Voos,M.,
    Chang,L.L., Esaki,L.
    Cyclotron resonance and far-infrared
    magneto-absorption experiments on semimetallic
    InAs-GaSb superlattices
    Phys. Rev. Lett. 45   1719-1722 (1980)

424 Harris,J.J., Woodcock,J.M.
    Low noice GaAs varactor and mixer diodes prepared by
    molecular beam epitaxy
    Electron. Lett. 16   317-319 (1980)

425 Haydl,W.H., Smith,R.S., Bosch,R.
    50-110 GHz Gunn diodes using molecular beam epitaxy
    IEEE Electron Device Lett. EDL-1   224-226 (1980)

426 Haydl,W.H., Smith,R.S., Bosch,R.
    100-GHz Gunn diodes fabricated by molecular beam
    epitaxy
    Appl. Phys. Lett. 37   556-557 (1980)

427 Heckingbottom,R., Davies,G.J.
T   Germanium doping of gallium arsenide grown by
       molecular beam epitaxy - some thermodynamic aspects
    J. Cryst. Growth 50    644-647 (1980)

428 Heckingbottom,R., Todd,C.J., Davies,G.J.
T   The interplay of thermodynamics and kinetics in
       molecular beam epitaxy of doped gallium arsenide
    J. Electrochem. Soc. 127    445-450 (1980)

429 Hiyamizu,S., Mimura,T., Fujii,T., Nanbu,K.
    High mobility of two-dimensional electrons at the
       GaAs/n-AlGaAs heterojunction interface
    Appl. Phys. Lett 37    805-807 (1980)

430 Hiyamizu,S., Nanbu,K., Sakurai,T., Hashimoto,H.,
       Fujii,T., Ryuzan,O.
    Lateral definition of monocrystalline GaAs prepared
       by molecular beam epitaxy
    J. Electrochem. Soc. 127    1562-1567 (1980)

431 Hollis,M., Dandekar,N., Eastman,L.F., Shur,M.S.,
       Woodard,D.W., Stall,R.A., Wood,C.E.C.
    Transverse magnetoresistance in GaAs two terminal
       submicron devices: a characterization of electron
       transport in the near ballistic regime
    IEDM 80 Technical Digest    622-625 (1980)

432 Kasamanyan,Z.H., Yuzbashyan,E.S.
T   Green's function of the heterojunction superlattice
    Phys. Status Solidi B 97    K149-K152 (1980)

433 Kawai,N.J., Chang,L.L., Sai-Halasz,G.A.,
       Chang,C.A., Esaki,L.
    Magnetic field-induced semimetal-to-semiconductor
       transition in InAs-GaSb superlattices
    Appl. Phys. Lett. 36    369-371 (1980)

434 Knodle,W.S., Luscher,P.E.
R   Recent developments in device fabrication by MBE
    Semicond. Int. Nov. 1980    39-52 (1980)

435 Kogelnik,H.
R   Devices for optical communications
    Inst. Phys. Conf. Ser. 53    1-28 (1980)

436 Kroemer,H., Polasko,K.J., Wright,S.L.
    On the (110) orientation as the preferred orientation
       for the molecular beam epitaxial growth of GaAs on
       Ge, GaP on Si, and similar zincblende-on-diamond
       systems
    Appl. Phys. Lett. 36    763-765 (1980)

437 **Kuebler,B., Ranke,W., Jacobi,K.**
LEED and AES of stoichiometric and arsenic-rich GaAs(110) surfaces prepared by molecular beam epitaxy
Surf. Sci. 92   519-527 (1980)

438 **Kuenzel,H., Fischer,A., Ploog,K.**
Quantitative evaluation of substrate temperature dependence of Ge incorporation in GaAs during molecular beam epitaxy
Appl. Phys. 22   23-30 (1980)

439 **Kuenzel,H., Ploog,K.**
The effect of As2 and As4 molecular beam species on photoluminescence of molecular beam epitaxially grown GaAs
Appl. Phys. Lett. 37   416-418 (1980)

440 **Lee,T.P., Burrus,C.A., Cho,A.Y.**
Low threshold current transverse junction lasers on semi-insulating substrates by MBE
Electron. Lett. 16   510-511 (1980)

441 **Lee,T.P., Burrus,C.A., Cho,A.Y., Cheng,K.Y., Manchon,D.D.,Jr.**
Zn-diffused back-illuminated p-i-n photodiodes in InGaAs/InP grown by molecular beam epitaxy
Appl. Phys. Lett. 37   730-731 (1980)

442 **Ludeke,R.**
R  Surface studies on MBE-grown III-V compounds and alloys
J. Vac. Sci. Technol. 17   1241-1246 (1980)

443 **Luscher,P.E., Knodle,W.S., Chai,Y.G.**
R  Automated molecular beams grow thin semiconducting films
Electronics August   160-168 (1980)

444 **Malik,R.J., AuCoin,T.R., Ross,R.L., Board,K., Wood,C.E.C., Eastman,L.F.**
Planar-doped barriers in GaAs by molecular beam epitaxy
Electron. Lett. 16   836-838 (1980)

445 **Malik,R.J., Board,K., Eastman,L.F., Wood,C.E.C., AuCoin,T.R., Ross,R.L., Savage,R.O.**
GaAs planar doped barriers by molecular beam epitaxy
IEDM 80, Technical Digest   456-459 (1980)

446 **Massies,J., Chaplart,J., Linh,N.T.**
Contact metal - semi-conducteur: recents developpements de l'etude concernant l'arseniure de gallium
Rev. Tech. Thomson-CSF 12   281-309 (1980)

447 **Massies,J., Dezaly,F., Linh,N.T.**
Effects of H2S adsorption on surface properties of GaAs(100) grown in situ by MBE
J. Vac. Sci. Technol. 17    1134-1140 (1980)

448 **Massies,J., Etienne,P., Dezaly,F., Linh,N.T.**
Stoichiometry effects on surface properties of GaAs(100) grown in situ by MBE
Surf. Sci. 99    121-131 (1980)

449 **Matsumoto,N., Kumabe,K.**
Amorphous GaAs films by molecular beam deposition
Jpn. J. Appl. Phys. 19    1583-1590 (1980)

450 **Meggitt,B.T., Parker,E.H.C., King,R.M., Grange,J.D.**
Electrical and structural properties of InAs layers on (100)GaAs substrates prepared by molecular beam deposition
J. Cryst. Growth 50    538-548 (1980)

451 **Mendez,E.E., Chang,C.A., Chang,L.L., Esaki,L., Pollak,F.H.**
Electroreflectance study of semiconductor superlattices
J. Phys. Soc. Japan 49 Suppl. A    1009-1012 (1980)

452 **Merlin,R., Colvard,C., Klein,M.V., Morkoc,H., Cho,A.Y.**
Raman scattering in superlattices: Anisotropy of polar phonons
Appl. Phys. Lett. 36    43-45 (1980)

453 **Merlin,R., Colvard,C., Klein,M.V., Morkoc,H., Gossard,A.C., Cho,A.Y.**
Phonon folding and anisotropy in GaAs-AlAs superlattices
J. Phys. Soc. Jpn. 49, Suppl. A    1025-1028 (1980)

454 **Metze,G.M., Levy,H.M., Woodard,D.W., Wood,C.E.C., Eastman,L.F.**
GaAs integrated circuits by selected-area molecular beam epitaxy
Appl. Phys. Lett. 37    628-630 (1980)

455 **Metze,G.M., Stall,R.A., Wood,C.E.C., Eastman,L.F.**
Dependence of the electrical characteristics of heavily Ge-doped GaAs on molecular beam epitaxy growth parameters
Appl. Phys. Lett. 37    165-167 (1980)

456 **Miller,D.L., Harris,J.S.,Jr.**
Molecular beam epitaxial GaAs heteroface solar cell grown on Ge
Appl. Phys. Lett. 37    1104-1106 (1980)

457 **Miller,R.C., Kleinman,D.A., Nordland,W.A., Gossard,A.C.**
Luminescence studies of optically pumped quantum wells in GaAS-Al$_x$Ga$_{1-x}$AS multilayer structures
Phys. Rev. B22 863-871 (1980)

458 **Mimura,T., Hiyamizu,S., Fujii,T., Nanbu,K.**
A new field-effect transistor with selectively doped GaAs/n-Al$_x$Ga$_{1-x}$As heterojunctions
Jpn. J. Appl. Phys. 19 L225-L227 (1980)

459 **Moench,W., Gant,H.**
Combined LEED, AES, and work function studies during the formation of Ge:GaAs(110) heterostructures
J. Vac. Sci. Technol. 17 1094-1100 (1980)

460 **Mori,S., Ando,T.**
T Electronic properties of a heavily-doped n-type GaAs-Ga(1-x)Al(x)As superlattice
Surf. Sci. 98 101-107 (1980)

461 **Mori,S., Ando,T.**
T Electronic properties of a semiconductor superlattice II. Low temperature mobility perpendicular to the superlattice
J. Phys. Soc. Jpn. 48 865-873 (1980)

462 **Morkoc,H., Cho,A.Y., Radice,C.**
Transport properties of Sn-doped Al$_x$Ga$_{1-x}$As grown by molecular beam epitaxy
J. Appl. Phys. 51 4882-4884 (1980)

463 **Morkoc,H., Hopkins,C.G., Evans,C.A.,Jr., Cho,A.Y.**
Chromium and tellurium redistribution in GaAs and Al0.3Ga0.7As grown by molecular beam epitaxy
J. Appl. Phys. 51 5986-5991 (1980)

464 **Morkoc,H., Witkowski,L.C., Drummond,T.J., Stanchak,C.M., Cho,A.Y., Streetman,B.G.**
Growth conditions to achieve mobility enhancement in Al$_x$Ga$_{1-x}$As-GaAs heterojunctions by MBE
Electron. Lett. 16 753-754 (1980)

465 **Neave,J.H., Blood,P., Joyce,B.A.**
A correlation between electron traps and growth processes in n-GaAs prepared by molecular beam epitaxy
Appl. Phys. Lett. 36 311-312 (1980)

466 **Norris,M.T.**
Substrate temperature related degradation mechanisms in molecular beam epitaxial InP
Appl. Phys. Lett. 36 282-283 (1980)

467  Norris,M.T.
     Composition control of Ga(x)In(1-x)As films grown by
        MBE onto InP substrates
     Appl. Phys. Lett. 36    833-835 (1980)

468  Oe,K., Ando,S., Sugiyama,K.
     RHEED Study of InSb films grown by molecular beam
        epitaxy
     Jpn. J. Appl. Phys. 19    L417-L420 (1980)

469  Ohno,H., Barnard,J.A., Wood,C.E.C., Eastman,L.F.
     Double heterostructure Ga0.47In0.53As MESFETs by MBE
     IEEE Electron Device Lett. EDL-1    154-155 (1980)

470  Ohno,H., Barnard,J.A., Wood,C.E.C., Rathbun,L.,
        Eastman,L.F.
     Double heterojunction GaInAs devices by MBE
     IEDM 80 Technical Digest    434-437 (1980)

471  Palmier,J.F., Ballini,Y.
  T  Near equilibrium mobility tensor in a semiconductor
        superlattice: scattering by acoustical phonons
     J. Physique Lett. 41    L-539-L-542 (1980)

472  Panish,M.B.
  R  Molecular beam epitaxy
     Science 208    916-922 (1980)

473  Panish,M.B.
     Molecular beam epitaxy of GaAs and InP with gas
        source for As and P
     J. Electrochem. Soc. 127    2729-2733 (1980)

474  Panish,M.B., Cho,A.Y.
  R  Molecular beam epitaxy
     IEEE Spectrum April 1980    18-23 (1980)

475  Panish,M.B., Cho,A.Y.
  R  Epitaxy by molecular beam. Molecular beams in a
        vacuum chamber produce devices starting from the
        atom
     Elettrotecnica 67    1057-1064 (1980)

476  Passner,A., Gibbs,H.M., Gossard,A.C., McCall,S.L.,
        Venkatesan,T.N.C., Wiegmann,W.
     Ultrashort laser: Lasing in MBE GaAs layer with
        perpendicular-to-film optical excitation and
        emission
     IEEE J. Quantum Electron. QE-16    1283-1285 (1980)

477  Pinczuk,A., Worlock,J.M., Stoermer,H.L., Dingle,R.,
        Wiegmann,W., Gossard,A.C.
     Intersubband spectroscopy of two dimensional electron
        gases: Coulomb interactions
     Solid State Commun. 36    43-46 (1980)

478 **Pinczuk,A., Worlock,J.M., Stoermer,H.L., Dingle,R., Wiegmann,W., Gossard,A.C.**
Light scattering spectroscopy of two-dimensional electron gases in semiconductors
J. Phys. Soc. Japan 49   Suppl. A, 1025-1028 (1980)

479 **Pinczuk,A., Worlock,J.M., Stoermer,H.L., Dingle,R., Wiegmann,W., Gossard,A.C.**
Inelastic light scattering spectroscopy of a multilayer two-dimensional electron gas
Surf. Sci. 98   126-133 (1980)

480 **Ploog,K.**
R  Molecular Beam epitaxy of III-V compounds
In "Crystals, Growth, Properties and Applications" Ed. H.C. Freyhardt (Springer, Berlin, Heidelberg, New York) Vol. 3   73-162 (1980)

481 **Pollmann,J.**
TR  On the electronic structure of semiconductor surfaces, interfaces, and defects at surfaces or interfaces
in "Festkoerperprobleme", Ed. J. Treusch (Vieweg, Braunschweig, 1980) Vol. XX   117-175 (1980)

482 **Sai-Halasz,G.A.**
R  Man-made semiconductor superlattices
Lect. Notes Phys. 122   215-225 (1980)

483 **Sakaki,H.**
T  Scattering suppression and high-mobility effect of size-quantized electrons in ultrafine semiconductor wire structures
Jpn. J. Appl. Phys. 19   L735-L738 (1980)

484 **Saris,F.W., Chu,W.K., Chang,C.A., Ludeke,R., Esaki,L.**
Ion backscattering and channeling study of InAs-GaSb superlattices
Appl. Phys. Lett. 37   931-933 (1980)

485 **Schorr,A.J., Tsang,W.T.**
Development of self-pulsations due to delf-annealing of proton bombarded regions during aging in proton bombarded stripe-geometry AlGaAs DH Lsaers grown by molecular beam epitaxy
IEEE J. Quantum Electron. QE-16   898-901 (1980)

486 **Scott,G.B., Roberts,J.S., Lee,R.F.**
Optically pumped laser action at 77 K in GaAs/GaInP double heterostructures grown by molecular beam epitaxy
Appl. Phys. Lett. 37   30-32 (1980)

487 Shur,M.S., Eastman,L.F.
T   Ballistic and near ballistic transport in GaAs
    IEEE Electron Device Lett. EDL-1   147-148 (1980)

488 Shur,M.S., Eastman,L.F.
T   GaAs n+-n--n+ ballistic structure
    Electron. Lett. 16   522-523 (1980)

489 Smith,R.S., Ganser,P.M., Hiesinger,P.,
        Koschel,W.H.
R   Duenne Galliumarsenid-Schichten durch
        Molekularstrahl-Epitaxie: Wachstumsbedingungen und
        physikalische Eigenschaften.
    Vakuum-Technik 28   231-238 (1980)

490 Stall,R.A., Wood,C.E.C., Kirchner,P.D.,
        Eastman,L.F.
    Growth-parameter dependence of deep levels in
        molecular-beam-epitaxial GaAs
    Electron. Lett. 16   171-172 (1980)

491 Stoermer,H.L.
R   Modulation doping of semiconductor superlattices and
        interfaces
    J. Phys. Soc. Jpn. 49 Suppl.A   1013-1020 (1980)

492 Stoermer,H.L., Tsang,W.T.
    Two-dimensional hole gas at a semiconductor
        heterojunction interface
    Appl. Phys. Lett. 36   685-687 (1980)

493 Tsang,W.T.
    Cw multiwavelength transverse-junction-stripe lasers
        grown by molecular beam epitaxy operating
        predominantly in single-longitudinal modes
    Appl. Phys. Lett. 36   441-443 (1980)

494 Tsang,W.T.
    Infrared-visible (0.89-0.72mum) $Al_xGa_{1-x}As/Al_yGa_{1-y}As$
        double-heterostructure lasers grown by molecular
        beam epitaxy
    J. Appl. Phys. 51   917-919 (1980)

495 Tsang,W.T.
    Symmetric separate confinement heterostructure lasers
        with low threshold and narrow beam divergence by
        M.B.E.
    Electron. Lett. 16   939-941 (1980)

496 Tsang,W.T.
    Al(x)Ga(1-x)As/Al(y)Ga(1-y)As heterostructure lasers
        grown by molecular beam epitaxy
    IEEE, Optical Soc. America, Top. Meet. Integrated and
        Guided-Wave Optics, Technical Digest   TuA1/1-5
        (1980)

**497** Tsang, W.T.
Very low current threshold GaAs-AlxGa1-xAs double-heterostructure lasers grown by molecular beam epitaxy
Appl. Phys. Lett. 36   11-14 (1980)

**498** Tsang, W.T., Hartman, R.L., Elder, H.E., Holbrook, W.R.
Cw electro-optical characteristics and preliminary reliability of double-heterostructure GaAs-AlxGa1-xAs lasers grown by molecular beam epitaxy
Appl. Phys. Lett. 37   141-143 (1980)

**499** Tsang, W.T., Logan, R.A.
GaAs-AlxGa1-xAs buried-heterostructure lasers grown by molecular beam epitaxy with Al0.65Ga0.53As(Ge-doped)liquid phase epitaxy overgrown layer for current injection confinement
Appl. Phys. Lett. 36   730-733 (1980)

**500** Tsang, W.T., Reinhart, F.K., Ditzenberger, J.A.
The effect of substrate temperature on the current threshold of GaAsAlxGa1-xAs double-heterostructure lasers grown by molecular beam epitaxy
Appl. Phys. Lett. 36   118-120 (1980)

**501** Tsui, D.C., Englert, T., Cho, A.Y., Gossard, A.C.
Observation of magnetophonon resonances in a two-dimensional electronic system
Phys. Rev. Lett. 44   341-344 (1980)

**502** Tsui, D.C., Stoermer, H.L., Gossard, A.C., Wiegmann, W.
Two-dimensional electrical transport in GaAs-AlxGa1-xAs multilayers at high magnetic fields
Phys. Rev. B21   1589-1595 (1980)

**503** Voisin, P., Guldner, Y., Vieren, J.P., Voos, M., Benoit, C., Kawai, N.J., Chang, L.L., Esaki, L.
Optical studies of InAs-GaSb superlattices
J. Phys. Soc. Jpn. 49, Suppl. A   1005-1008 (1980)

**504** Witkowski, L.C., Drummond, T.J., Stanchak, C.M., Morkoc, H.
High mobilities in AlxGa1-xAs-GaAs heterojunctions
Appl. Phys. Lett. 37   1033-1035 (1980)

**505** Wood, C.E.C.
R   Progress, problems, and application of molecular-beam epitaxy
in "Physics of Thin Films", Vol. 11, Eds. W.R. Hunter and G. Haas (Academic Press, New York, 1980) 35-103 (1980)

506  Wood,C.E.C., DeSimone,D., Judaprawira,S.
     Improved molecular-beam epitaxial GaAs power FET's
     J. Appl. Phys.  51   2074-2078 (1980)

507  Wood,C.E.C., Metze,G.M., Berry,J.D., Eastman,L.F.
     Complex free-carrier profile synthesis by
        "atomic-plane" doping of MBE GaAs
     J. Appl. Phys. 51   383-387 (1980)

508  Zehr,S.W., Yang,H.T., Coleman,J.J., Miller,D.L.,
        Yang,J.J.J., Ruth,R.P., Harris,J.S.,Jr.
  R  Monolithic multicolor solar conversion
     Proc. SPIE - Soc. Photo-Opt. Instr. Eng. 248
        125-129 (1980)

# Subject Categories and References
# Year 1981

### AlAs, GaAs, (AlGa)As
509,510,511,512,513,514,515,516,517,518,519,522,523,524,
525,526,527,530,531,532,533,534,535,536,537,538,539,541,
542,544,545,546,547,550,551,554,556,557,558,559,560,561,
562,563,564,565,566,567,568,569,570,571,572,573,574,577,
578,582,583,585,587,588,589,590,591,592,593,594,595,596,
597,598,599,600,601,602,603,605,606,609,610,611,612,613,
614,615,616,617,618,619,620,621,622,623,624,625,626,627,
628,629,630,633,634,635,636,637,638,639,640,641,642,643,
644,645,646,650,651,652,653,655,659,660,661,664,665,666,
667,668,669,670,671,672,673,674,675,676,677,678,679,680,
681,682,685,686,687,688,689,690,691,693,694,695,696,697,
698,699,700,701,703,704,705,706,708,709,710,711,712,713,
714,715,716,717,718,719,720,721,722,723,726,727,728,729,
730,731,732,733,734,735,736,737,738,739,740,741,742,743

### GaSb, InAs, Ga(AsSb), (GaIn)As
521,528,529,530,531,540,548,549,552,553,554,555,557,558,
573,575,576,583,607,630,631,636,647,648,649,657,658,668,
671,672,692,693,694,695,707,709,724,725,735,741,742,743

### InP, (AlIn)As, (GaIn)As, (GaIn)(PAs)
520,521,528,529,540,552,553,554,555,573,575,577,578,583,
607,630,636,647,648,649,657,658,662,663,671,683,692,695,
702,707,709,741,742,743

### Growth Equipment
520,525,536,537,552,553,555,556,575,583,586,605,607,615,
624,630,638,645,663,670,671,672,673,675,697,711,723

### Growth Mechanism, Surface Analysis
511,512,513,514,515,516,523,524,525,533,537,555,577,579,
580,582,583,587,588,599,605,612,616,620,622,624,625,634,
638,651,654,656,658,659,668,671,675,681,693,694,695,696,
702,704,726,733,736,740,742

### Periodic Multilayer Structure
509,512,517,518,530,531,536,537,543,544,547,548,549,557,
563,564,565,566,575,576,578,581,584,585,590,597,598,601,
631,636,639,640,642,646,661,667,668,669,670,672,673,674,
675,676,686,688,695,697,710,713,714,716,724,725,730,731,
732,733,734

## Modulation Doping
509,519,527,532,537,561,562,565,566,567,568,569,570,578,
585,590,591,592,594,595,596,600,602,603,606,610,611,623,
626,627,643,644,650,652,653,661,669,672,673,677,678,679,
682,697,698,699,700,705,721,722,728,730,739

## Electrical Properties
510,517,518,519,520,521,522,526,527,528,529,530,531,532,
533,535,537,538,539,542,545,546,548,550,551,552,554,556,
559,560,561,562,563,565,566,567,568,569,570,572,573,574,
575,578,580,581,583,585,586,588,590,592,593,594,595,596,
600,602,603,606,607,608,609,610,611,613,614,615,617,618,
619,621,623,626,627,628,629,632,633,634,635,637,643,644,
648,650,652,653,655,656,657,658,659,660,662,664,669,670,
671,672,673,674,675,676,677,679,680,682,683,685,688,696,
698,699,700,701,702,705,706,707,721,722,723,725,727,728,
729,730,739,741,743

## Optical Properties
509,510,511,512,513,514,518,520,521,528,530,531,537,539,
543,549,550,551,556,557,558,563,564,571,572,575,583,585,
597,598,601,602,607,608,613,615,620,628,631,636,639,640,
641,642,658,665,667,668,669,671,672,673,674,675,683,684,
686,687,689,690,691,699,701,703,706,707,708,709,710,711,
712,713,714,715,716,717,718,719,720,724,725,729,730,731,
732,733,734,735,737

## Microwave Devices
522,528,534,536,537,561,562,573,578,589,591,594,595,602,
605,606,614,627,629,630,637,643,644,655,657,660,678,722,
727,741

## Optoelectronic Devices
521,528,535,536,537,539,550,551,602,605,630,665,706,707,
708,709,710,711,712,713,714,715,716,717,718,719,729

509 Abstreiter,G., Zeller,C., Ploog,K.
Study of GaAs-Al(x)Ga(1-x)As multilayer systems by resonant inelastic light scattering techniques
Inst. Phys. Conf. Ser. 56   741-749 (1981)

510 Afsar,M.N., Button,K.J., Cho,A.Y., Morkoc,H.
Ultimate method for unambiguous identification of all donors in epitaxial GaAs and related compounds
Int. J. Infrared Millimeter Waves 2   1113-1121 (1981)

511 Alferov,Z.I., Ber,B.Y., Kop'ev,P.S., Mel'tser,B.Y., Minchev,G.M.
Synthesis of GaAs and Al(x)Ga(1-x)As films by molecular-beam epitaxy
Sov. Tech. Phys. Lett. 6   1-2   *   Pis'ma Zh. Tekh. Fiz. 7   3-5 (1981)

512 Alvarado,S.F., Ciccacci,F., Campagna,M.
GaAs-AlxGa1-xAs superlattices as sources of polarized photoelectrons
Appl. Phys. Lett. 39   615-617 (1981)

513 Alvarado,S.F., Ciccacci,F., Valeri,S., Campagna,M., Feder,R., Pleyer,H.
Spin polarized photoemission from molecular beam epitaxy-grown Be-doped GaAs
Z. Phys. B 44   259-264 (1981)

514 Alvarado,S.F., Hopster,H., Feder,R., Pleyer,H.
T Spin-orbit effects in LEED form Ni(001): experiment and theory
Solid State Commun. 39   1319-1322 (1981)

515 Andersson,T.G., Svensson,S.P.
Temperature dependence of electron loss spectra from MBE-grown GaAs(001)
Surf. Sci. 110   L578-L582 (1981)

516 Andersson,T.G., Svensson,S.P.
The initial growth of Au on GaAs(001)-c(4X4)
Surf. Sci. 110   L583-L586 (1981)

517 Ando,T.
T Magnetic Quantization and transport in a semiconductor superlattice
in "Physics in High Magnetic Fields", Eds. S. Chikazumi and M. Miura, Springer Ser. Solid State Sci. Vol. 24   301-304 (1981)

518 Ando,T.
T Electronic properties of a semiconductor superlattice.III.Energy levels and transport in magnetic fields
J. Phys. Soc. Jpn. 50   2978-2984 (1981)

519 Aoki,H., Ando,T.
T  Effect of localization on the Hall conductivity in
      the two-dimensional system in strong magnetic
      fields
   Solid State Commun. 38   1079-1082 (1981)

520 Asahi,H., Kawamura,Y., Ikeda,M., Okamoto,H.
   Molecular-beam epitaxial growth of InP homoepitaxial
      layers and their electrical and optical properties
   J. Appl. Phys. 52   2852-2859 (1981)

521 Asahi,H., Kawamura,Y., Ikeda,M., Okamoto,H.
   Near room temperature CW operation at 1.70 um of MBE
      grown InGaAs/InP DH lasers
   Jpn. J. Appl. Phys. 20   L187-L190 (1981)

522 Asbeck,P.M., Miller,D.L., Milano,R.A.,
      Harris,J.S.,Jr., Kaelin,G.R., Zucca,R.
   (Ga,Al)As/GaAs bipolar transistors for digital
      integrated circuits
   IEDM 81 Technical Digest   629-632 (1981)

523 Bachrach,R.Z., Bauer,R.S., Chiaradia,P.,
      Hansson,G.V.
   Surface phases of GaAs(100) and AlAs(100)
   J. Vac. Sci. Technol. 18   797-801 (1981)

524 Bachrach,R.Z., Bauer,R.S., Chiaradia,P.,
      Hansson,G.V.
   Reconstructions of GaAs and AlAs surfaces as a
      function of metal to As ratio
   J. Vac. Sci. Technol. 19   335-343 (1981)

525 Bachrach,R.Z., Krusor,B.S.
   Morphological defects arising during MBE growth of
      GaAs
   J. Vac. Sci. Technol. 18   756-764 (1981)

526 Ballingall,J.M., Stall,R.A., Wood,C.E.C.,
      Eastman,L.F.
   Electron transport across the aprupt Ge-GaAs n-n
      heterojunction
   J. Appl. Phys. 52   4098-4103 (1981)

527 Baraff,G.A., Tsui,D.C.
   Explanation of quantized-Hall-resistance plateaus in
      heterojunction inversion layers
   Phys. Rev. B 24   2274-2277 (1981)

528 Barnard,J.A., Ohno,H., Wood,C.E.C., Eastman,L.F.
   Integrated double heterostructure Ga(0.47)In(0.53)As
      photoreceiver with automatic gain control
   IEEE Electron Device Lett. EDL-2   7-9 (1981)

529 Barnard,J.A., Wood,C.E.C., Eastman,L.F.
Resistivity increase in MBE $Ga_{0.47}In_{0.53}As$ following ion bombardment
IEEE Electron Device Lett. EDL-2   193-195 (1981)

530 Bastard,G.
T Hydrogenic impurity states in a quantum well: A simple model
Phys. Rev. B 24   4714-4722 (1981)

531 Bastard,G.
T Superlattice band structure in the envelope-function approximation
Phys. Rev. B 24   5693-5697 (1981)

532 Basu,P.K., Nag,B.R.
T Piezoelectric scattering in quantised surface layers in semiconductors
J. Phys. C 14   1519-1522 (1981)

533 Blanchet,R.C., Delhomme,B.J., Urgell,J.J.
Schottky diodes showing the effects of various in situ GaAs treatments in molecular beam epitaxy (MBE)
Inst. Phys. Conf. Ser. 56   613-620 (1981)

534 Board,K., Chandra,A., Wood,C.E.C., Judaprawira,S., Eastman,L.F.
T Characteristics of planar doped FET structures
IEEE Trans. Electron Devices ED-28   505-510 (1981)

535 Bouchaib,P., Contour,J.P., Raymond,F., Verie,C., D'Avitaya,F.A.
Use of molecular beam epitaxy for the achievement of low resistance intercell contacts in multiband gap solar cells
J. Vac. Sci. Technol. 19   145-147 (1981)

536 Brenac,A.
R L'epitaxie par jets moleculaires
Rev. Polytech. (Suisse) 8bis No.1414   1144-1147 (1981)

537 Brenac,A.
R L'epitaxie par jets moleculaires
Recherches 104   3-14 (1981)

538 Calawa,A.R.
On the use of $AsH_3$ in the molecular beam epitaxial growth of GaAs
Appl. Phys. Lett. 38   701-703 (1981)

539 Capasso,F., Logan,R.A., Tsang,W.T., Hayes,J.R.
Channeling photodiode: A new versatile interdigitated p-n junction photodetector
Appl. Phys. Lett. 41   944-946 (1981)

540 Carter,C.B., Wood,C.E.C., Roberts,J.S.
Glide of dissociated dislocations in III-V compounds
Appl. Phys. Lett. 38   805-807 (1981)

541 Chai,Y.G., Chow,R.
Source and elimination of oval defects on GaAs films grown by molecular beam epitaxy
Appl. Phys. Lett. 38   796-798 (1981)

542 Chai,Y.G., Chow,R., Wood,C.E.C.
The effect of growth conditions on Si incorporation in molecular beam epitaxial GaAs
Appl. Phys. Lett. 39   800-803 (1981)

543 Chaikovskii,I.A., Shmelev,G.M., Enaki,N.A.
T Light absorption in semiconductors with superlattice in quantizing magnetic field
Phys.Status Solidi B 108   559-566 (1981)

544 Chang,C.A., Chu,W.K., Mendez,E.E., Chang,L.L., Esaki,L.
Molecular beam epitaxy of Ge-GaAs superlattices
J. Vac. Sci. Technol. 19   567-570 (1981)

545 Chang,C.A., Heiblum,M., Ludeke,R., Nathan,M.I.
Effect of substrate surface treatment in molecular beam epitaxy on the vertical electronic transport through the film-substrate interface
Appl. Phys. Lett. 39   229-231 (1981)

546 Chang,C.A., Heiblum,M., Nathan,M.I.
Monolayer-Sn technique for solving the substrate-film interface high-resistance problem in molecular beam epitaxy
IBM Tech. Disclosure Bull. 24   1969 (1981)

547 Chang,C.A., Segmueller,A., Chang,L.L., Esaki,L.
Ge-GaAs superlattices by molecular beam epitaxy
Appl. Phys. Lett. 38   912-914 (1981)

548 Chang,L.L., Kawai,N.J., Mendez,E.E., Chang,C.A., Esaki,L.
Semimetallic InAs-GaSb superlattices to the heterojunction limit
Appl. Phys. Lett. 38   30-32 (1981)

549 Chang,L.L., Sai-Halasz,G.A., Esaki,L., Aggarwal,R.L.
Spatial separation of carriers in InAs-GaSb superlattices
J. Vac. Sci. Technol. 19   589-591 (1981)

550 Chen,C.Y., Cho,A.Y., Garbinski,P.A., Bethea,C.G.
An ultahigh speed modulated barrier photodiode made on p-type gallium arsenide substrates
IEEE Electron Device Lett. EDL-2   290-292 (1981)

551  Chen,C.Y., Cho,A.Y., Garbinski,P.A., Bethea,C.G., Levine,B.F.
     Modulated barrier photodiode: A new majority-carrier photodetector
     Appl. Phys. Lett. 39   340-342 (1981)

552  Cheng,K.Y., Cho,A.Y., Bonner,W.A.
     Beryllium doping in Ga0.47In0.53As and Al0.48In0.52As grown by molecular beam epitaxy
     J. Appl. Phys. 52   4672-4675 (1981)

553  Cheng,K.Y., Cho,A.Y., Wagner,W.R.
     Molecular beam epitaxial growth of uniform Ga0.47In0.53As with a rotating sample holder
     Appl. Phys. Lett. 39   607-609 (1981)

554  Cheng,K.Y., Cho,A.Y., Wagner,W.R.
     Tin doping in Ga0.47In0.53As and Al0.48In0.52As grown by molecular beam epitaxy
     J. Appl. Phys. 52   6328-6330 (1981)

555  Cheng,K.Y., Cho,A.Y., Wagner,W.R., Bonner,W.A.
     Molecular beam epitaxial growth of uniform In0.53Ga0.47As on InP with a coaxial In-Ga oven
     J. Appl. Phys. 52   1015-1021 (1981)

556  Cho,A.Y., Cheng,K.Y.
     Growth of extremely uniform layers by rotating substrate holder with molecular beam epitaxy for applications to electro-optic and microwave devices
     Appl. Phys. Lett. 38   360-362 (1981)

557  Combescot,M., Benoit a la Guillaume,C.
 T   Two-dimensional electrons-holes droplets in superlattices
     Solid State Commun. 39   651-654 (1981)

558  Das Sarma,S., Madhukar,A.
 T   Collective modes of spatially separated, two-component, two-dimensional plasma in solids
     Phys. Rev. B 23   805-815 (1981)

559  Davies,G.J., Andrews,D.A., Heckingbottom,R.
     Electrochemical sulfur doping of GaAs grown by molecular beam epitaxy
     J. Appl. Phys. 52   7214-7218 (1981)

560  Day,D.S., Oberstar,J.D., Drummond,T.J., Morkoc,H., Cho,A.Y., Streetman,B.G.
     Electron traps created by high temperature annealing in MBE n-GaAs
     J. Electron. Mater. 10   445-453 (1981)

561 Delagebeaudeuf,D., Linh,N.T.
T   Charge control of the heterojunction two-dimensional
       electron gas for MESFET application
    IEEE Trans. Electron Devices ED-28   790-795 (1981)

562 Delescluse,P., Laviron,M., Chaplart,J.,
       Delagebeaudeuf,D., Linh,N.T.
    Transport properties in GaAs-Al(x)Ga(1-x)As
       heterostructures and MESFET application
    Electron. Lett. 17   342-344 (1981)

563 Doehler,G.H.
R   Semiconductor superlattices − a new material for
       research and applications
    Phys. Scr. 24   430-439 (1981)

564 Doehler,G.H., Kuenzel,H., Olego,D., Ploog,K.,
       Ruden,P., Stolz,H.J., Abstreiter,G.
    Observation of tunable band gap and two-dimensional
       subbands in a novel GaAs superlattice
    Phys. Rev. Lett. 47   864-867 (1981)

565 Drummond,T.J., Keever,M., Kopp,W., Morkoc,H.,
       Hess,K., Streetman,B.G., Cho,A.Y.
    Field dependence of mobility in Al0.2Ga0.8As/GaAs
       heterojunctions at very low fields
    Electron. Lett. 17   545-547 (1981)

566 Drummond,T.J., Kopp,W., Morkoc,H.
    Three period (Al,Ga)As/GaAs heterostructures with
       extremely high mobilities
    Electron. Lett. 17   442-444 (1981)

567 Drummond,T.J., Kopp,W., Morkoc,H., Hess,K.,
       Cho,A.Y., Streetman,B.G.
    Effect of background doping on the electron mobility
       of (Al,Ga)As/GaAs heterostructures
    J. Appl. Phys. 52   5689-5690 (1981)

568 Drummond,T.J., Morkoc,H., Cho,A.Y.
    Dependence of electron mobility on spatial separation
       of electrons and donors in $Al_xGa_{1-x}As$/GaAs
       heterostructures
    J. Appl. Phys. 52   1380-1386 (1981)

569 Drummond,T.J., Morkoc,H., Hess,K., Cho,A.Y.
    Experimental and theoretical electron mobility of
       modulation doped $Al_xGa_{1-x}As$/GaAs heterostructures
       grown by molecular beam epitaxy
    J. Appl. Phys. 52   5231-5234 (1981)

570 Drummond,T.J., Morkoc,H., Su,S.L., Fischer,R.,
       Cho,A.Y.
    Enhanced mobility in inverted $Al_xGa_{1-x}As$/GaAs
       heterojunctions: Binary on top of ternary
    Electron. Lett. 17   870-871 (1981)

571   Duggan,G., Scott,G.B., Foxon,C.T., Harris,J.J.
      Photoluminescence technique for the determination of
         minority-carrier diffusion length in GaAs grown by
         molecular beam epitaxy
      Appl. Phys. Lett. 38   246-248 (1981)

572   Duhamel,N., Henoc,P., Alexandre,F., Rao,E.V.K.
      Influence of growth temperature on Be incorporation
         in molecular beam epitaxy GaAs epilayers
      Appl. Phys. Lett. 39   49-51 (1981)

573   Eastman,L.F.
      Experimental studies of ballistic transport in
         semiconductors
      J. Physique 42 Colloque C 7   C7/263-C7/269 (1981)

574   Eastman,L.F., Stall,R.A., Woodard,D.W.,
         Wood,C.E.C., Dandekar,N., Shur,M.S., Hollis,M.,
         Board,K.
      Ballistic electron transport in thin layers of GaAs
      Inst. Phys. Conf. Ser. 56   185-192 (1981)

575   Esaki,L.
  R   InAs-GaSb superlattices - synthesized semiconductors
         and semimetals
      J. Cryst. Growth 52   227-240 (1981)

576   Esaki,L., Chang,L.L., Mendez,E.E.
  T   Polytype superlattices and multi-heterojunctions
      Jpn. J. Appl. Phys. 20   L529-L532 (1981)

577   Farrow,R.F.C.
  R   The use of ion beams in molecular beam epitaxy
      Thin Solid Films 80   197-211 (1981)

578   Farrow,R.F.C.
  R   Recent european developments in MBE
      J. Vac. Sci. Technol. 19   150-156 (1981)

579   Farrow,R.F.C., Jones,G.R., Williams,G.M.,
         Young,I.M.
      Molecular beam epitaxial growth of high structural
         perfection, heteroepitaxial CdTe films on InP(001)
      Appl. Phys. Lett. 39   954-956 (1981)

580   Farrow,R.F.C., Sullivan,P.W., Williams,G.M.,
         Jones,G.R., Cameron,D.C.
      MBE-grown fluoride films: A new class of epitaxial
         dielectrics
      J. Vac. Sci. Technol. 19   415-420 (1981)

581   Ferry,D.K.
  T   Charge instabilities in lateral super-lattices under
         conditions of population inversion
      Phys. Status Solidi B 106   63-71 (1981)

582 **Foxon,C.T.**
R  Molecular beam epitaxy - surface and kinetic effects
CRC Crit. Rev. Solid State & Mater. Sci. 10   235-242 (1981)

583 **Foxon,C.T., Joyce,B.A.**
R  Fundamental aspects of molecular beam epitaxy
in "Current Topics in Materials Science", Ed. E. Kaldis, Vol. 7 (North-Holland, Amsterdam/New York, 1981)   1-68 (1981)

584 **Gaponov,S.V., Luskin,B.M., Salashchenko,N.N.**
Homoepitaxial superlattices with nonoriented barrier layers
Solid State Commun. 39   301-302 (1981)

585 **Gornik,E., Schawarz,R., Tsui,D.C., Gossard,A.C., Wiegmann,W.**
Far infrared emission from 2D electrons at the GaAs-Al$_x$Ga$_{1-x}$As interface
Solid State Commun. 38   541-545 (1981)

586 **Gotoh,H., Suga,T., Suzuki,H., Kimata,M.**
Low temperature growth of gallium nitride
Jpn. J. Appl. Phys. 20   L545-L548 (1981)

587 **Harris,J.J., Joyce,B.A.**
Comments on "RED intensity oscillations during MBE of GaAs"
Surf. Sci. 108   L444-L446 (1981)

588 **Harris,J.J., Joyce,B.A., Dobson,P.J.**
Oscillations in the surface structure of Sn-doped GaAs during growth by MBE
Surf. Sci. 103   L90-L96 (1981)

589 **Haydl,W.H.**
Harmonic operation of GaAs millimetre wave transferred electron oscillators
Electron. Lett. 17   825-826 (1981)

590 **Hess,K.**
R  Lateral transport in superlattices
J. Physique 42 Colloque C7   C7/3-C7/17 (1981)

591 **Hikosaka,K., Mimura,T., Joshin,K.**
Selective dry etching of AlGaAs-GaAs heterojunction
Jpn. J. Appl. Phys. 20   L847-L850 (1981)

592 **Hiyamizu,S., Fujii,T., Mimura,T., Nanbu,K., Saito,J., Hashimoto,H.**
The effect of growth temperature on the mobility of two-dimensional electron gas in selectively doped GaAs/N-AlGaAs heterostructures grown by MBE
Jpn. J. Appl. Phys. 20   L455-L458 (1981)

593  Hiyamizu,S., Fujii,T., Nanbu,K., Hashimoto,H.
    Lateral uniformity in Sn- or Si-doped n-GaAs grown by molecular beam epitaxy
    J. Cryst. Growth 51   149-152 (1981)

594  Hiyamizu,S., Mimura,T.
 R   MBE-grown selectively doped GaAs/N-AlGaAs heterostructures and their application to high electron mobility transistors
    in "Semiconductor Technologies" Ed. J. Mishizawa (OHMSHA, Tokyo, 1981)   258-271 (1981)

595  Hiyamizu,S., Mimura,T., Fujii,T., Nanbu,K., Hashimoto,H.
    Extremely high mobility of two dimensional electron gas in selectively doped GaAs/N-AlGaAs heterojunction structures grown by MBE
    Jpn. J. Appl. Phys. 20   L245-L248 (1981)

596  Hiyamizu,S., Nanbu,K., Mimura,T., Fujii,T., Hashimoto,H.
    Room-temperature mobility of two-dimensional electron gas selectively doped GaAs/N-AlGaAs heterojunction structures
    Jpn. J. Appl. Phys. 20   L378-L380 (1981)

597  Holonyak,N.,Jr., Laidig,W.D., Camras,M.D., Morkoc,H., Drummond,T.J., Hess,K.
    Clustering and phonon effects in $Al_xGa_{1-x}As$-GaAs quantum-well heterostructure lasers grown by molecular beam epitaxy
    Solid State Commun. 40   71-74 (1981)

598  Holonyak,N.,Jr., Laidig,W.D., Camras,M.D., Morkoc,H., Drummond,T.J., Hess,K., Burroughs,M.S.
    Clustering in molecular beam epitaxial $Al_xGa_{1-x}As$-GaAs quantum-well heterostructure lasers
    J. Appl. Phys. 52   7201-7207 (1981)

599  Ihm,J., Joannopoulos,J.D.
 T   Ground-state properties of GaAs and AlAs
    Phys. Rev. B 24   4191-4197 (1981)

600  Inoue,M., Hiyamizu,S., Hida,H., Hashimoto,H., Inuishi,Y.
    Hot electron effects in a 2D electron gas at the GaAs/AlGaAs interface
    J. Physique 42 Colloque C7   C7/19-C7/24 (1981)

601  Ishibashi,T., Suzuki,Y., Okamoto,H.
    Photoluminescence of an AlAs/GaAs superlattice grown by MBE in the 0.7-0.8 um wavelength region
    Jpn. J. Appl. Phys. 20   L623-L626 (1981)

602 Ishikawa,H., Mimura,T., Hiyamizu,S.
R   Opening in molecular beam epitaxy devices
    in "Solid State Devices 1981, Proc. ESSDERC-11 &
       SSSDT-6" (Les Ulis. Les Editions de Physique,
       Paris, 1981)    137-152 (1981)

603 Ishikawa,T., Hiyamizu,S., Mimura,T., Saito,J.,
       Hashimoto,H.
    The effect of annealing on the electrical properties
       of selectivly doped GaAs/N-AlGaAs Heterojunction
       structures grown by MBE
    Jpn. J. Appl. Phys. 20    L814-L816 (1981)

604 Johannessen,J.S., Clegg,J.B., Foxon,C.T.,
       Joyce,B.A.
    Interface composition profiles of MBE grown GaP films
       on GaAs substrates
    Phys. Scr. 24    440-443 (1981)

605 Joyce,B.A.
R   Molecular beam epitaxy
    Phys. Educ. 16    328-332 (1981)

606 Judaprawira,S., Wang,W.I., Chao,P.C., Wood,C.E.C.,
       Woodard,D.W., Eastman,L.F.
    Modulation-doped MBE GaAs/n-Al$_x$Ga$_{1-x}$As MESFETs
    IEEE Electron Device Lett. EDL-2    14-15 (1981)

607 Kawamura,Y., Asahi,H., Ikeda,M., Okamoto,H.
    Improved properties of In(x)Ga(1-x)As layers grown by
       molecular-beam epitaxy on InP substrates
    J. Appl. Phys. 52    3445-3452 (1981)

608 Kawamura,Y., Asahi,H., Nagai,H.
    Molecular beam epitaxial growth of undoped
       low-resistivity In(x)Ga(1-x)P on GaAs at high
       substrate temperatures (500-580 C)
    Jpn. J. Appl. Phys. 20    L807-L810 (1981)

609 Kazarinov,R.F., Luryi,S.
T   Charge injection over triangular barriers in unipolar
       semiconductor structures
    Appl. Phys. Lett. 38    810-812 (1981)

610 Keever,M., Drummond,T.J., Hess,K., Morkoc,H.,
       Streetman,B.G.
    Time dependence of current at high electric fields in
       Al$_x$Ga$_{1-x}$As-GaAs heterojunction layers
    Electron. Lett. 17    93-94 (1981)

611 Keever,M., Shichijo,H., Hess,K., Banerjee,S.,
       Witkowski,L., Morkoc,H., Streetman,B.G.
    Measurements of hot-electron conduction and
       real-space transfer in GaAs-Al$_x$Ga$_{1-x}$As
       heterojunction layers
    Appl. Phys. Lett. 38    36-38 (1981)

612 Kirchner,P.D., Woodall,J.M., Freeouf,J.L.,
    Pettit,G.D.
    Volatile metal-oxide incorporation in layers of
      GaAs,Ga(1-x)Al(x)As and related compounds grown by
      molecular beam epitaxy
    Appl. Phys. Lett. 38   427-429 (1981)

613 Kirchner,P.D., Woodall,J.M., Freeouf,J.L.,
    Wolford,D.J., Pettit,G.D.
    Volatile metal oxide incorporation in layers of GaAs
      and Ga(1-x)Al(x)As grown by molecular beam epitaxy
    J. Vac. Sci. Technol. 19   604-606 (1981)

614 Kondoh,H., Berenz,J., Hierl,T.L., Dalman,G.C.,
    Lee,C.A.
    High efficiency mode characterization in a 20 GHz MBE
      GaAs IMPATT diode amplifier
    IEEE MTT-S Int. Microwave Symposium Digest   238-240
      (1981)

615 Koschel,W.H., Smith,R.S., Hiesinger,P.
    Optical and electrical characterization of chemical
      defects in GaAs layers grown by MBE
    J. Electrochem. Soc. 128   1336-1338 (1981)

616 Kowalczyk,S.P., Miller,D.L., Waldrop,J.R.,
    Grant,R.W., Newman,P.G.
    Protection of molecular beam epitaxy grown AlxGa1-xAs
      epilayers during ambient transfer
    J. Vac. Sci. Technol. 19   255-256 (1981)

617 Kroemer,H.
  T Analytic approximations for degenerate accumulation
      layers in semiconductors, with applications to
      barrier lowering in isotype heterojunctions
    J. Appl. Phys. 52   873-878 (1981)

618 Kuenzel,H., Doehler,G.H., Fischer,A., Ploog,K.
    Modulation of two-dimensional conductivity in a
      molecular beam epitaxially grown GaAs bulk
      space-charge system
    Appl. Phys. Lett. 38   171-174 (1981)

619 Kuenzel,H., Graf,K., Hafendoerfer,M., Fischer,A.,
    Ploog,K.
  R IEC-Bus-Mess-System fuer die Halbleiterforschung: 1)
      Charakteristische Messgroessen an
      Schottky-Kontakten   2) Aufbau von Hard- und
      Software   3)Messergebnisse und Diskussion
    Techn. Messen 48   295-300, 397-401, 435-440 (1981)

620 Kuenzel,H., Ploog,K.
    Sharp-line luminescence transitions due to growth
      induced point defects in MBE GaAs
    Inst. Phys. Conf. Ser. 56   519-528 (1981)

621 Kunc,K., Martin,R.M.
T   Atomic structure and properties of polar Ge-GaAs(100)
       interfaces
    Phys. Rev. B 24    3445-3455 (1981)

622 Landgren,G., Ludeke,R.
    Al-reactions with GaAs (100) surfaces
    Solid State Commun. 37    127-131 (1981)

623 Landwehr,G.
R   Praezisionsbestimmung der Feinstrukturkonstanten aus
       Magnetotransportmessungen an
       Halbleiter-Randschichten
    Phys. Bl. 37    59-65 (1981)

624 Larsen,P.K., Neave,J.H., Joyce,B.A.
    Angle-resolved photoemission from As-stable
       GaAs(001)surfaces prepared by MBE
    J. Phys. C 14    167-192 (1981)

625 Larsen,P.K., Veen,J.F. van der, Mazur,A.,
       Pollmann,J., Verbeek,B.H.
    Photoemission from vallence bands of GaAs(001) grown
       by molecular beam epitaxy
    Solid State Commun. 40    459-462 (1981)

626 Laughlin,R.B.
T   Quantized Hall conductivity in two dimensions
    Phys. Rev. B 23    5632-5633 (1981)

627 Laviron,M., Delagebeaudeuf,D., Delescluse,P.,
       Chaplart,J., Linh,N.T.
    Low-noise two-dimensional electron gas FET
    Electron. Lett. 17    536-537 (1981)

628 Lee,T.P., Holden,W.S., Cho,A.Y.
    Improved molecular beam epitaxial growth of
       AlxGa1-xAs/GaAs high-radiance LED's for optical
       communications
    IEEE J. Quantum Electron. QE-17    387-391 (1981)

629 Levy,H.M., Metze,G.M., Woodard,D.W., Camp,W.O.,
       Tiberio,R.C., Wood,C.E.C., Eastman,L.F.
    GaAs integrated circuits by selective epitaxy and
       electron beam lithography
    Solid State Technol. August    127-130 (1981)

630 Luscher,P.E.
R   Molecular beam epitaxy: An emerging epitaxy technology
    Thin Solid Films 83    125-141 (1981)

631 Maan,J.C., Guldner,Y., Vieren,J.P., Voisin,P.,
       Voos,M., Chang,L.L., Esaki,L.
    Three-dimensional character of semimetallic InAs-GaSb
       superlattices
    Solid State Commun. 39    683-686 (1981)

632 **Madhukar,A., Delgado,J.**
T  The electronic structure of Si/GaP(110) interface and superlattice
   Solid State Commun. 37   199-203 (1981)

633 **Malik,R.J., Board,K., Eastman,L.F., Wood,C.E.C., AuCoin,T.R., Ross,R.L.**
   Rectifying, variable planar-doped-barrier structures in GaAs
   Inst. Phys. Conf. Ser. 56   697-710 (1981)

634 **Massies,J., Chaplart,J., Laviron,M., Linh,N.T.**
   Monocrystalline aluminium ohmic contact to n-GaAs by H2S adsorption
   Appl. Phys. Lett. 38   693-695 (1981)

635 **McAfee,S.R., Tsang,W.T., Lang,D.V.**
   The effect of substrate growth temperature on deep levels in n-$Al_xGa_{1-x}As$ grown by molecular beam epitaxy
   J. Appl. Phys. 52   6165-6167 (1981)

636 **Mendez,E.E., Chang,L.L., Landgren,G., Ludeke,R., Esaki,L., Pollak,F.H.**
   Observation of superlattice effects on the electronic bands of multilayer heterostructures
   Phys. Rev. Lett. 46   1230-1234 (1981)

637 **Metze,G.M., Levy,H.M., Woodard,D.W., Wood,C.E.C., Eastman,L.F.**
   GaAs integrated circuits by selective molecular beam epitaxy
   Inst. Phys. Conf. Ser. 56   161-170 (1981)

638 **Miller,D.L., Newman,P.G.**
   Low cost spinner for semiconductor surface preparation prior to MBE growth
   J. Vac. Sci. Technol. 19   124 (1981)

639 **Miller,R.C., Kleinman,D.A., Munteanu,O., Tsang,W.T.**
   New transitions in the photoluminescence of GaAs quantum wells
   Appl. Phys. Lett. 39   1-3 (1981)

640 **Miller,R.C., Kleinman,D.A., Tsang,W.T., Gossard,A.C.**
   Observation of the excited level of excitons in GaAs quantum wells
   Phys. Rev. B24   1134-1136 (1981)

641 **Miller,R.C., Tsang,W.T.**
   Al-Ga disorder in $Al_xGa_{1-x}As$ alloys grown by molecular beam epitaxy
   Appl. Phys. Lett. 39   334-335 (1981)

642 Miller,R.C., Weisbuch,C., Gossard,A.C.
Alloy clustering in $Al_xGa_{1-x}As$
Phys. Rev. Lett. 46   1042 (1981)

643 Mimura,T., Hiyamizu,S., Joshin,K., Hikosaka,K.
Enhancement-mode high electron mobility transistors for logic applications
Jpn. J. Appl. Phys. 20   L317-L319 (1981)

644 Mimura,T., Joshin,K., Hiyamizu,S., Hikosaka,K., Abe,M.
High electron mobility transistor logic
Jpn. J. Appl. Phys. 20   L598-L600 (1981)

645 Minchev,G.M., Ber,B.Y., Kop'ev,P.S., Mel'tser,B.Y.
Effective substrate preparation method for molecular-ion epitaxy
Sov. Tech. Phys. Lett. 7   518-519   *   Pis'ma Zh. Tekh. Fiz. 7   1209-1213 (1981)

646 Mon,K.K., Hess,K., Dow,J.D.
T  Deformation potentials of superlattices and interfaces
J. Vac. Sci. Technol. 19   564-566 (1981)

647 Morgan,D.V., Ohno,H., Wood,C.E.C., Eastman,L.F., Berry,J.D.
Ion beam analysis of molecular beam epitaxy InAlAs/InGaAs layer structures
J. Electrochem. Soc. 128   2419-2424 (1981)

648 Morgan,D.V., Ohno,H., Wood,C.E.C., Schaff,W.J., Board,K., Eastman,L.F.
Characterisation of Al/AlInAs/GaInAs heterostructures
IEE Proc. 128,Pt. I   141-143 (1981)

649 Morgan,D.V., Wood,C.E.C., Ohno,H., Eastman,L.F.
Channeling analysis of MBE InAlAs/InGaAs interfaces
J. Vac. Sci. Technol. 19   596-598 (1981)

650 Morkoc,H.
Current transport in modulation doped (Al,Ga)As/GaAs heterostructures: Applications to high speed FET's
IEEE Electron Device Lett. EDL-2   260-262 (1981)

651 Munoz-Yague,A., Piqueras,J., Fabre,N.
Preparation of carbon-free GaAs surfaces: AES and RHEED analysis
J. Electrochem. Soc. 128   149-153 (1981)

652 Narita,S., Takeyama,S., Luo,W., Hiyamizu,S., Nanbu,K., Hashimoto,H.
Galvanomagnetic study of 2-dimensional electron gas in $Al_xGa_{1-x}As$/GaAs heterojunction FET
Jpn. J. Appl. Phys. 20   L443-L446 (1981)

653  Narita,S., Takeyama,S., Luo,W., Hiyamizu,S.,
     Nanbu,K., Hashimoto,H.
     Magnetoconductance investigations of AlxGa1-xAs/GaAs
        heterojunction FET in strong magnetic fields
     Jpn. J. Appl. Phys. 20   L447-L450 (1981)

654  Noreika,A.J., Francombe,M.H., Wood,C.E.C.
     Growth of Sb and InSb by molecular-beam epitaxy
     J. Appl. Phys. 52   7416-7420 (1981)

655  O'Clock,G.D.,Jr., Erickson,L.P., Mattord,T.J.
  R  MBE: Precise processing for better EHF devices
     Microwaves,August    101-105 (1981)

656  Oe,K., Ando,S., Sugiyama,K.
     InSb(1-x)Bi(x) films grown by molecular beam epitaxy
     Jpn. J. Appl. Phys. 20   L303-L306 (1981)

657  Ohno,H., Barnard,J.A., Rathbun,L., Wood,C.E.C.,
     Eastman,L.F.
     Double heterostructure Ga(0.47)In(0.53)As MESFETs by
        molecular beam epitaxy
     Inst. Phys. Conf. Ser. 56   465-473 (1981)

658  Ohno,H., Wood,C.E.C., Rathbun,L., Morgan,D.V.,
     Wicks,G.W., Eastman,L.F.
     GaInAs-AlInAs structures grown by molecular beam
        epitaxy
     J. Appl. Phys. 52   4033-4037 (1981)

659  Okamoto,K., Wood,C.E.C., Eastman,L.F.
     Schottky barrier heights of molecular beam epitaxial
        metal-AlGaAs structures
     Appl. Phys. Lett. 38   636-638 (1981)

660  Omori,M., Drummond,T.J., Morkoc,H.
     Low-noice GaAs field-effect transistors prepared by
        molecular beam epitaxy
     Appl. Phys. Lett. 39   566-569 (1981)

661  Palmier,J.F.
  R  Les super-reseaux artificiels
     Echo Rech. 105    41-48 (1981)

662  Park,R.M., Stanley,C.R.
     Characterisation of a deep electron trap in
        molecular-beam epitaxial InP
     Electron. Lett. 17   669-670 (1981)

663  Park,R.M., Stanley,C.R., Clampitt,R.
     A low-energy ion source for p-type doping in MBE
     Inst. Phys. Conf. Ser. 54   235-240 (1981)

664  Parker,E.H.C., Kubiak,R.A., King,R.M., Grange,J.D.
     An investigation into silicon doping of MBE(100) GaAs
     J. Phys. D 14   1853-1865 (1981)

665  Pawlik,J.R., Tsang,W.T., Nash,F.R., Hartman,R.L., Swaminathan,V.
Reduced temperature dependence of threshold of (Al,Ga)As lasers grown by molecular beam epitaxy
Appl. Phys. Lett. 38  974-976 (1981)

666  Petroff,P.M., Feldman,L.C., Cho,A.Y., Williams,R.S.
Properties of aluminium epitaxial growth on GaAs
J. Appl. Phys. 52  7317-7320 (1981)

667  Petroff,P.M., Weisbuch,C., Dingle,R., Gossard,A.C., Wiegmann,W.
Luminescence properties of GaAs-Ga$_{1-x}$Al$_x$As double heterostructures and multiquantum-well superlattices grown by molecular beam epitaxy
Appl. Phys. Lett. 38  965-967 (1981)

668  Phillips,J.C.
T  Electronic and atomic structure of type-II semiconductor superlattices
Phys. Rev. B 24  3620-3622 (1981)

669  Pinczuk,A., Worlock,J.M., Stoermer,H.L., Gossard,A.C., Wiegmann,W.
Light scattering spectroscopy of electrons in GaAs-(AlGa)As heterostructures: Correlation with transport properties
J. Vac. Sci. Technol. 19  561-563 (1981)

670  Piqueras,J.
R  Crecimiento epitaxial por haces moleculares
Mundo Electronico 107  45-50 (1981)

671  Ploog,K.
R  Molecular beam epitaxy of III-V compounds: technology and growth process
Ann. Rev. Mater. Sci. 11  171-210 (1981)

672  Ploog,K.
R  Halbleiterschichtstrukturen aus dem System GaAs-AlAs mit neuartigen Eigenschaften: 1) Epitaxie durch Molekularstrahlung  2) Schichtdicken von 10 Atomabstaenden  3) FETs mit extrem kurzen Schaltzeiten
Markt + Technik Nr.27 77-79, Nr. 29 74-77, Nr. 31 70-71 (1981)

673  Ploog,K., Fischer,A.
Herstellung von GaAs/Al(x)Ga(1-x)As-Mehrschichtstrukturen mit definierten optischen und elektrischen Eigenschaften mittels Molekularstrahl-Epitaxie
Forschungsbericht BMFT-FB-T81-042 1-68 (1981)

674  Ploog,K., Fischer,A., Doehler,G.H., Kuenzel,H.
Novel periodic doping structures in GaAs grown by molecular beam epitaxy
Inst. Phys. Conf. Ser. 56   721-730 (1981)

675  Ploog,K., Fischer,A., Kuenzel,H.
The use of Si and Be impurities for novel periodic doping structures in GaAs grown by molecular beam epitaxy
J. Electrochem. Soc. 128   400-410 (1981)

676  Ploog,K., Kuenzel,H., Knecht,J., Fischer,A., Doehler,G.H.
Simultaneous modulation of electron and hole conductivity in a new periodic GaAs doping multilayer structure
Appl. Phys. Lett. 38   870-873 (1981)

677  Prange,R.E.
T  Quantized Hall resistance and the measurement of the fine-structure constant
Phys. Rev. B 23   4802-4805 (1981)

678  Price,P.J.
T  Mesostructure electronics
IEEE Trans. Electron Devices ED-28   911-914 (1981)

679  Price,P.J.
T  Two-dimensional electron transport in semiconductor layers II: screening
J. Vac. Sci. Technol. 19   599-603 (1981)

680  Price,P.J.
T  Two-dimensional electron transport in semiconductor layers. 1) Phonon Scattering
Ann. Phys. 133   217-239 (1981)

681  Ranke,W., Jacobi,K.
R  Structure and reactivity of GaAs surfaces
Prog. Surf. Sci. 10   1-52 (1981)

682  Rendell,R.W., Girvin,S.M.
T  Hall voltage dependence on inversion-layer geometry in the quantum Hall-effect regime
Phys. Rev. B 23   6610-6614 (1981)

683  Roberts,J.S., Dawson,P., Scott,G.B.
Homoepitaxial molecular beam growth of InP on thermally cleaned (100) oriented substrates
Appl. Phys. Lett. 38   905-907 (1981)

684  Roberts,J.S., Scott,G.B., Gowers,J.P.
Structural and photoluminescent properties of $Ga_xIn_{1-x}P$ (x=0.5) grown on GaAs by molecular beam epitaxy
J. Appl. Phys. 52   4018-4026 (1981)

685 **Sakaki,H., Sekiguchi,Y., Sun,D.C., Taniguchi,M., Ohno,H., Tanaka,A.**
Schottky-barrier properties of nearly-ideal (n=1) Al contacts on MBE- and heat cleaned-GaAs surfaces
Jpn. J. Appl. Phys. 20   L107-L110 (1981)

686 **Satpathy,S., Altarelli,M.**
T  Model calculation of the optical properties of semiconductor quantum wells
Phys. Rev. B 23   2977-2982 (1981)

687 **Schulman,J.N., Chang,Y.C.**
T  New method for calculating electronic properties of superlattices using complex band structures
Phys. Rev. B 24   4445-4448 (1981)

688 **Schulman,J.N., McGill,T.C.**
T  Complex band structure and superlattice electronic states
Phys. Rev. B 23   4149-4155 (1981)

689 **Schwartz,G.P., Cho,A.Y.**
Chemical reaction at the Al-GaAs interface
J. Vac. Sci. Technol. 19   607-610 (1981)

690 **Scott,G.B., Duggan,G., Dawson,P.**
A photoluminescence study of beryllium-doped GaAs grown by molecular beam epitaxy
J. Appl. Phys. 52   6888-6894 (1981)

691 **Scott,G.B., Duggan,G., Roberts,J.S.**
Interface recombination velocity and misfit strain in molecular-beam epitaxy double heterostructures of $GaAs/Ga_xIn_{1-x}P (0.47 < x < 0.51)$
J. Appl. Phys. 52   6312-6315 (1981)

692 **Serrano,C.M., Chang,C.A.**
Reduction in dislocation densities in the step-graded growth of InGaAs by molecular-beam epitaxy
Appl. Phys. Lett. 39   808-809 (1981)

693 **Skeath,P., Lindau,I., Su,C.Y., Spicer,W.E.**
Models of column III and V elements on GaAs(110): application to MBE
J. Vac. Sci. Technol. 19   556-560 (1981)

694 **Skeath,P., Su,C.Y., Lindau,I., Spicer,W.E.**
Bonding of column 3 and 5 atoms on GaAs(110)
Solid State Commun. 40   873-876 (1981)

695 **Srobar,F.**
T  Configuration entropy of ternary alloy superlattice
Cryst. Res. Technol. 16   1173-1180 (1981)

696 Stall,R.A., Wood,C.E.C., Board,K., Dandekar,N.,
    Eastman,L.F., Devlin,W.J.
    A study of Ge/GaAs interfaces grown by molecular beam
       epitaxy
    J. Appl. Phys. 52   4062-4069 (1981)

697 Stanchak,C.M., Morkoc,H., Witkowski,L.C.,
    Drummond,T.J.
    Automatic shutter controller for molecular beam
       epitaxy
    Rev. Sci. Instrum. 52   438-442 (1981)

698 Stoermer,H.L., Gossard,A.C., Wiegmann,W.
    Backside-gated modulation-doped GaAs-(AlGa)As
       heterojunction interface
    Appl. Phys. Lett. 39   493-495 (1981)

699 Stoermer,H.L., Gossard,A.C., Wiegmann,W.,
    Baldwin,K.
    Dependence of electron mobility in modulation-doped
       GaAs-(AlGa)As heterojunction interfaces on
       electron density and Al concentration
    Appl. Phys. Lett. 39   912-914 (1981)

700 Stoermer,H.L., Pinczuk,A., Gossard,A.C.,
    Wiegmann,W.
    Influence of an undoped (AlGa)As spacer on mobility
       enhancement in GaAs-(AlGa)As superlattices
    Appl. Phys. Lett. 38   691-693 (1981)

701 Stringfellow,G.B., Stall,R.A., Koschel,W.H.
    Carbon in molecular beam epitaxial GaAs
    Appl. Phys. Lett. 38   156-157 (1981)

702 Sullivan,P.W., Farrow,R.F.C., Jones,G.R.,
    Stanley,C.R.
    MBE growth of InP and epitaxial aluminium contacts
    Inst. Phys. Conf. Ser. 56   45-54 (1981)

703 Swaminathan,V., Tsang,W.T.
    Effect of growth temperature on the photoluminescent
       spectra from Sn-doped $Ga_{1-x}As$ grown by molecular
       beam epitaxy
    Appl. Phys. Lett. 38   347-349 (1981)

704 Taylor,J.A.
    An XPS study of the oxidation of AlAs thin films
       grown by MBE
    J. Vac. Sci. Technol. 20   751-755 (1981)

705 Thouless,D.J.
 T  Localisation and the two-dimensional Hall effect
    J. Phys. C 14   3475-3480 (1981)

**706  Tsang,W.T.**
R   Reliable (AlGa)As DH lasers grown by molecular beam
       epitaxy for optical communication systems
    Proc. SPIE - Int. Soc. Opt. Eng. 269   17-24 (1981)

**707  Tsang,W.T.**
    Al0.48In0.52As/Ga0.47In0.53As/Al0.48In0.52As
       double-heterostructure lasers grown by molecular
       beam epitaxy with lasing wavelength at 1.65mum
    J. Appl. Phys. 52   3861-3864 (1981)

**708  Tsang,W.T.**
    A new current-injection heterostructure laser:the
       double-barrier double-heterostructure laser
    Appl. Phys. Lett. 38   835-837 (1981)

**709  Tsang,W.T.**
    Extension of lasing wavelenghts beyond 0.87mum in
       GaAs/AlxGa1-xAs double-heterostructure lasers by
       In incorporation in the GaAs active layers during
       molecular beam epitaxy
    Appl. Phys. Lett. 38   661-663 (1981)

**710  Tsang,W.T.**
    Extremely low threshold (AlGa)As modified
       multiquantum well heterostructure lasers grown by
       molecular beam epitaxy
    Appl. Phys. Lett. 39   786-788 (1981)

**711  Tsang,W.T.**
    High-through-put, high-yield, and highly-reproducible
       (AlGa)As double-heterostructure laser wafers grown
       by molecular beam epitaxy
    Appl. Phys. Lett. 38   587-589 (1981)

**712  Tsang,W.T.**
    A graded-index waveguide separate-confinement laser
       with very low threshold and a narrow Gaussian beam
    Appl. Phys. Lett. 39   134-137 (1981)

**713  Tsang,W.T.**
R   Novel optoelectronic devices prepared by molecular
       beam epitaxy
    Proc. SPIE - Int. Soc. Opt. Eng. 317   66-73 (1981)

**714  Tsang,W.T.**
    Device characteristics of !alGa)As multiquantum-well
       heterostructure lasers grown by molecular beam
       epitaxy
    Appl. Phys. Lett. 58   204-207 (1981)

**715  Tsang,W.T., Ditzenberger,J.A.**
    A visible (AlGa)As heterostructure laser grown by
       molecular beam epitaxy
    Appl. Phys. Lett. 39   193-194 (1981)

716 **Tsang,W.T., Hartman,R.L.**
cw narrow beam (AlGa)As multiquantum-well
   heterostructure lasers grown by molecular beam
   epitaxy
Appl. Phys. Lett. 38    502-504 (1981)

717 **Tsang,W.T., Hartman,R.L., Schwartz,B., Fraley,P.E.,
   Holbrook,W.R.**
The reliability of (AlGa\As double-hetrostucture
   lasers grown by molecular beam epitaxy
Appl. Phys. Lett. 39    683-685 (1981)

718 **Tsang,W.T., Holbrook,W.R., Fraley,P.E.**
The high-temperature (55-70C) device characteristics
   of cw (AlGa)As double-heterostructure
   proton-bombarded stripe lasers grown by molecular
   beam epitaxy
Appl. Phys. Lett. 38    6-9 (1981)

719 **Tsang,W.T., Holbrook,W.R., Fraley,P.E.**
Optical self-pulsation behavior of cw (AlGa)As
   shallow proton-bombarded and narrow-striped (5mum)
   double-heterostructure lasers grown by molecular
   beam epitaxy
Appl. Phys. Lett. 39    34-37 (1981)

720 **Tsang,W.T., Swaminathan,V.**
The effect of As/Ga flux ratio on the
   photoluminescent spectra from molecular beam
   epitaxially-grown Sn-doped $Al_xGa_{1-x}As$
Appl. Phys. Lett. 39    486-487 (1981)

721 **Tsui,D.C., Gossard,A.C.**
Resistance standard using quantization of the Hall
   resistance of $GaAs-Al_xGa_{1-x}As$ heterostructures
Appl. Phys. Lett. 38    550-552 (1981)

722 **Tsui,D.C., Gossard,A.C., Kaminsky,G., Wiegmann,W.**
Transport properties of $GaAs-Al_xGa_{1-x}As$
   heterojunction field-effect transistors
Appl. Phys. Lett. 39    712-714 (1981)

723 **Veuhoff,E., Pletschen,W., Balk,P., Lueth,H.**
Metalorganic CVD of GaAs in a molecular beam system
J. Cryst. Growth 55    30-34 (1981)

724 **Voisin,P., Bastard,G., Goncalves da Silva,C.E.T.,
   Voos,M., Chang,L.L., Esaki,L.**
Luminescence from InAs-GaSb superlattices
Solid State Commun. 39    79-82 (1981)

725 **Voos,M., Esaki,L.**
R    InAs-GaSb Superlattices in high magnetic fields
   in "Physics in High Magnetic Fields", Eds.
      S.Chikazumi and N.Miura, Springer Ser. Solid State
      Sci. Vol. 24    292-300 (1981)

726 Waldrop,J.R., Kowalczyk,S.P., Grant,R.W.,
Kraut,E.A., Miller,D.L.
XPS measurement of GaAs-AlAs heterojunction band
discontinuities: Growth sequence dependence
J. Vac. Sci. Technol. 19   573-575 (1981)

727 Wang,W.I., Judaprawira,S., Wood,C.E.C.,
Eastman,L.F.
Molecular beam epitaxial GaAs-Al$_x$Ga$_{1-x}$As
heterostructures for metal semiconductor field
effect transistor applications
Appl. Phys. Lett. 38   708-710 (1981)

728 Wang,W.I., Wood,C.E.C., Eastman,L.F.
Extremely high electron mobilities in
modulation-doped GaAs-Al$_x$Ga$_{1-x}$As heterojunction
superlattices
Electron. Lett. 17   36-37 (1981)

729 Weimann,G., Schlapp,W., Burkhard,H.
MBE growth of GaAs an GaAlAs for the fabrication of
DH lasers
Phys. Status Solidi A 64   K99-K103 (1981)

730 Weisbuch,C.
R Les superreseaux des cristaux contre nature
Recherche 12   No. 118   100-102 (1981)

731 Weisbuch,C., Dingle,R., Gossard,A.C., Wiegmann,W.
Optical properties and interface disorder of
GaAs-Al(x)Ga(1-x)As multi-quantum well structures
Inst. Phys. Conf. Ser. 56   711-720 (1981)

732 Weisbuch,C., Dingle,R., Gossard,A.C., Wiegmann,W.
Optical characerization of interface disorder in
GaAs-Ga$_{1-x}$Al$_x$As multi-quantum well structures
Solid State Commun. 38   709-712 (1981)

733 Weisbuch,C., Dingle,R., Petroff,P.M., Gossard,A.C.,
Wiegmann,W.
Dependance of the structural and optical properties
of GaAs-Ga$_{1-x}$Al$_x$As multiquantum-well structures on
growth temperature
Appl. Phys. Lett. 38   840-842 (1981)

734 Weisbuch,C., Miller,R.C., Dingle,R., Gossard,A.C.,
Wiegmann,W.
Intrinsic radiative recombination from quantum states
in GaAs-Al$_x$Ga$_{1-x}$As multi-quantum well structures
Solid State Commun. 37   219-222 (1981)

735 White,S.R., Sham,L.J.
T Electronic properties of flat-band semiconductor
heterostructures
Phys. Rev. Lett. 47   879-882 (1981)

736 **Whitehouse,S.B., Foxon,C.T., Joyce,B.A.**
Thermal desorption spectroscopy of condensed lead films on (100) GaAs surfaces
Appl. Phys. A 26  27-33 (1981)

737 **Wicks,G.W., Wang,W.I., Wood,C.E.C., Eastman,L.F., Rathbun,L.**
Photoluminescence of $Al_xGa_{1-x}As$ grown by molecular beam epitaxy
J. Appl. Phys. 52  5792-5796 (1981)

738 **Williams,R.S., Feldman,L.C., Cho,A.Y.**
Channeling at the crystal-crystal interface: Al on GaAs(001)
Radiat. Effects 54  217-220 (1981)

739 **Witkowski,L.C., Drummond,T.J., Barnett,S.A., Morkoc,H., Cho,A.Y., Greene,J.E.**
High mobility $GaAs-Al_xGa_{1-x}As$ single period modulation-doped heterojunctions
Electron. Lett. 17  126-128 (1981)

740 **Wood,C.E.C.**
RED intensity oscillations during MBE of GaAs
Surf. Sci. 108  L441-L443 (1981)

741 **Wood,C.E.C.**
R Novel device structures by molecular beam epitaxy
J. Vac. Sci. Technol. 18  772-777 (1981)

742 **Wood,C.E.C., Rathbun,L., Ohno,H., DeSimone,D.**
On the origin and elimination of macroscopic defects in MBE films
J. Cryst. Growth 51  299-303 (1981)

743 **Woodall,J.M., Freeouf,J.L., Pettit,G.D., Jackson,T.N., Kirchner,P.D.**
Ohmic contacts to n-GaAs using graded band gap layers of $Ga_{(1-x)}In_{(x)}As$ grown by molecular beam epitaxy
J. Vac. Sci. Technol. 19  626-627 (1981)

# Subject Categories and References
## Year 1982

**AlAs, GaAs, (AlGa)As**
744,745,746,747,748,749,750,751,752,753,754,755,756,757,
758,759,760,761,765,766,767,768,769,770,771,776,778,779,
780,781,783,784,785,786,787,788,789,791,792,793,796,797,
798,799,800,801,802,803,804,805,807,808,809,810,815,816,
818,821,823,824,829,830,831,832,833,835,836,837,838,839,
840,841,842,844,845,846,847,848,849,850,851,852,853,854,
855,856,857,858,859,860,861,862,863,864,865,866,867,868,
870,871,872,873,874,875,876,877,878,879,880,883,884,885,
886,887,889,890,891,893,894,895,896,897,898,899,900,901,
903,904,906,907,908,909,910,911,912,913,914,915,916,917,
918,919,920,922,923,924,925,926,927,928,929,930,931,932,
933,934,935,936,937,938,939,940,941,942,943,944,945,947,
948,949,953,954,955,956,958,959,960,961,962,963,964,965,
966,968,969,970,971,972,973,974,975,976,977,978,979,980,
981,982,983,984,987,989,990,991,992,993,994,995,996,997,
998,999,1000,1001,1002,1006,1007,1008,1010,1011,1013,
1014,1015,1016,1017,1018,1019,1020,1021,1022,1023,1024,
1025,1027,1028,1029,1030,1031,1032,1033,1034,1035,1036,
1037,1038,1039,1040,1042,1043,1047,1048,1050,1052,1053,
1054,1055,1057,1058,1059,1060,1061,1062,1063,1064,1065,
1066,1067,1068,1069,1070,1071,1073,1074,1075,1076,1077,
1078,1079,1081,1082,1083,1084,1085,1086,1087,1088,1089,
1090,1092,1093,1095,1096,1097,1098,1099,1100,1101,1102,
1107,1108,1109,1112,1113,1114,1116,1117,1118,1119,1120,
1121,1122,1123,1124,1125,1126,1127,1128,1129,1130,1134,
1135,1136,1137,1138,1139,1140,1141,1142,1144,1145,1146,
1147,1148,1149,1152,1153,1155,1156,1157,1159,1162,1163,
1164,1165,1169,1170,1171,1172

**GaSb, InAs, Ga(AsSb), (GaIn)As**
757,758,771,773,774,775,776,777,778,786,794,795,799,806,
810,812,813,814,817,819,820,822,825,826,827,828,830,834,
846,869,881,882,888,895,905,942,946,951,965,986,988,989,
1003,1009,1026,1041,1049,1051,1068,1072,1077,1110,1111,
1115,1143,1148,1150,1151,1154,1157

**InP, (AlIn)As, (GaIn)As, (GaIn)(PAs)**
757,771,772,786,795,799,806,810,814,817,819,820,822,825,
826,827,828,830,869,895,904,921,942,946,951,1003,1004,
1005,1026,1049,1051,1056,1072,1104,1106,1110,1111,1115,
1131,1132,1133,1150,1154,1157

## Growth Equipment
762,839,883,886,888,898,903,912,913,917,927,947,967,970,
1046,1047,1064,1070,1131,1142,1149,1161,1167

## Growth Mechanism, Surface Analysis
748,750,751,752,753,767,769,770,780,781,782,789,797,807,
808,809,810,811,812,824,828,829,835,844,847,855,870,883,
888,891,898,902,908,912,926,940,941,942,945,948,952,965,
967,970,972,973,974,975,984,998,999,1000,1001,1002,1004,
1024,1033,1037,1038,1040,1044,1054,1055,1065,1071,1074,
1082,1087,1094,1104,1105,1106,1112,1140,1142,1147,1152,
1154,1155,1158,1160,1166,1167

## Periodic Multilayer Structure
746,757,758,761,773,774,775,776,778,784,791,792,793,794,
798,799,802,803,804,806,810,813,832,834,837,838,845,846,
857,858,862,880,881,882,894,895,897,898,905,909,910,919,
929,935,937,943,955,957,968,971,985,986,989,991,1008,
1012,1013,1014,1019,1020,1021,1025,1030,1035,1041,1050,
1057,1058,1063,1064,1066,1067,1068,1069,1070,1077,1078,
1079,1081,1085,1086,1090,1091,1092,1108,1115,1117,1127,
1128,1139,1143,1150,1159,1165,1169,1170,1172

## Modulation Doping
744,745,754,755,756,757,760,783,792,816,817,818,820,823,
826,837,838,845,846,849,850,851,852,853,860,861,862,863,
864,865,866,867,869,870,871,872,877,878,879,880,886,890,
897,898,906,915,916,917,918,920,925,930,931,932,946,953,
954,955,956,959,962,976,977,979,980,981,987,1011,1023,
1028,1031,1032,1039,1042,1048,1052,1053,1057,1058,1059,
1060,1061,1062,1064,1068,1072,1073,1075,1076,1077,1084,
1089,1096,1097,1098,1099,1100,1113,1114,1118,1119,1121,
1122,1123,1124,1134,1135,1136,1137,1138,1141,1147,1149,
1169

## Electrical Properties
744,745,747,748,749,755,756,757,758,759,760,762,764,765,
766,770,772,775,779,780,783,784,785,786,787,788,789,790,
793,794,795,796,799,800,801,803,806,811,812,813,814,815,
816,817,818,819,820,821,822,823,825,826,827,830,831,832,
836,837,838,839,840,841,842,847,848,849,851,852,853,854,
858,859,860,861,862,863,864,865,866,867,868,869,870,871,
872,873,875,876,877,878,879,880,881,882,885,890,892,896,
897,898,899,900,902,903,904,906,908,911,913,914,915,916,
917,918,920,921,925,928,930,931,932,933,934,936,938,940,
941,942,944,946,949,950,951,952,953,954,955,956,958,959,
960,961,962,963,964,966,968,969,970,976,977,979,980,981,
982,983,984,985,991,993,994,995,996,997,1003,1005,1006,
1007,1008,1009,1011,1015,1016,1017,1018,1023,1024,1026,
1027,1028,1029,1031,1032,1034,1036,1038,1042,1043,1045,
1047,1048,1049,1051,1052,1053,1054,1059,1060,1061,1062,
1063,1068,1069,1070,1072,1073,1075,1076,1077,1078,1079,
1080,1081,1082,1084,1086,1089,1091,1093,1095,1096,1097,
1098,1099,1100,1101,1107,1110,1111,1112,1113,1114,1118,
1119,1120,1121,1122,1130,1131,1132,1134,1135,1136,1137,
1138,1142,1144,1146,1147,1148,1152,1153,1155,1156,1160,
1162,1168,1170,1171,1172

## Optical Properties
746,748,750,754,757,761,762,763,764,770,771,772,775,776,
778,784,785,787,791,792,793,794,796,798,799,803,804,806,
816,818,821,828,830,832,835,845,846,847,854,856,857,858,
864,870,874,882,887,889,893,894,895,896,897,898,901,903,
905,907,909,910,920,921,922,923,935,937,938,939,940,943,
944,947,949,950,951,958,960,968,969,970,971,978,982,983,
985,986,987,988,990,992,1004,1005,1008,1010,1014,1019,
1020,1021,1022,1025,1030,1035,1041,1044,1045,1047,1050,
1056,1057,1058,1066,1067,1068,1069,1070,1071,1072,1077,
1088,1090,1093,1102,1103,1108,1109,1116,1117,1123,1124,
1125,1126,1127,1128,1130,1131,1132,1133,1141,1145,1146,
1154,1155,1159,1162,1163,1164,1165,1167,1168,1169,1170,
1171,1172

## Microwave Devices
744,745,765,766,772,779,783,814,817,820,822,823,830,831,
833,837,842,849,850,851,852,853,867,871,872,873,875,877,
885,904,913,915,916,917,918,925,928,961,962,963,964,977,
979,980,981,993,994,995,996,997,1007,1011,1015,1016,
1023,1027,1028,1029,1034,1036,1039,1047,1049,1051,1077,
1080,1084,1085,1100,1101,1110,1111,1119,1120,1137,1138,
1144,1153

## Optoelectronic Devices
748,761,763,764,771,785,786,798,799,801,802,803,804,816,
818,821,830,857,870,874,889,898,922,937,951,1017,1018,
1030,1035,1047,1050,1068,1095,1102,1125,1126,1127,1128,
1129,1130,1132,1133,1144,1145,1146,1150,1164,1165,1172

744 Abe,M., Mimura,T., Yokoyama,N., Ishikawa,H.
R  New technology towards GaAs LSI/VLSI for computer
      applications
   IEEE Trans. Microwave Theory Techn. MTT-30   992-998
      (1982)

745 Abe,M., Mimura,T., Yokoyama,N., Ishikawa,H.
R  New technology towards GaAs LSI/VLSI for computer
      applications
   IEEE Trans. Electron. Devices ED-29   1088-1094 (1982)

746 Abstreiter,G., Doehler,G.H., Kuenzel,H., Olego,D.,
      Ploog,K., Ruden,P., Stolz,H.J.
   Quantization of photoexcited electrons in GaAs nipi
      crystals
   Surf. Sci. 113   479-480 (1982)

747 Adachi,S., Kawashima,M., Kumabe,K., Yokoyama,K.,
      Tomizawa,M.
   Electron transport in GaAs n+/p-/n+ submicron diodes
   IEEE Electron Device Lett. EDL-3   409-411 (1982)

748 Alexandre,F., Duhamel,N., Ossart,P., Masson,J.M.,
      Meillerat,C.
   Problem related to the MBE growth at high substrate
      temperature for GaAs-Ga(1-x)Al(x)As double
      heterostructure lasers
   J. Physique 43 Colloque C5   C5/483-C5/489 (1982)

749 Alexandre,F., Masson,J.M., Post,G., Scavennec,A.
   AlN/GaAs structures grown by molecular beam epitaxy
      for metal/insulator/semiconductor devices
   Thin Solid Films 98   75-80 (1982)

750 Alvarado,S.F., Campagna,M., Hopster,H.
   Surface magnetism of Ni(001) near the critical region
      by spin-polarized electron scattering
   Phys. Rev. Lett. 48   51-54 (1982)

751 Andersson,T.G.
R  The initial growth of vapour deposited gold films
   Gold Bull. 15   7-18 (1982)

752 Andersson,T.G., Landgren,G., Svensson,S.P.
   Nucleation of Al on GaAs(001)
   Collected Papers of MBE-CST-2, Ed. R. Ueda (Jpn. Soc.
      Appl. Phys., Tokyo, 1982)   283-286 (1982)

753 Andersson,T.G., Svensson,S.P., Landgren,G.
   Interdiffusion and nucleation of Al monolayers on
      GaAs(001)-c(2x8) studied by AES
   J. Phys. C 15   6673-6676 (1982)

**754 Ando,T.**
T Self-consistent results for a GaAs/Al(x)Ga(1-x)As heterojunction. 1. Subband structure and light-scattering spectra
J. Phys. Soc. Jpn. 51    3893-3899 (1982)

**755 Ando,T.**
T Self-consistent results for a GaAs/Al(x)Ga(1-x)As heterojunction. 2. Low temperature mobility
J. Phys. Soc. Jpn.    3900-3907 (1982)

**756 Ando,T.**
T Hall effect and electron localization in a two-dimensional system in strong magnetic fields
in "Anderson Localization", Eds. Y. Nagaoka and H. Fukuyama, Springer Ser. Solid-State Sci. Vol. 39 176-190 (1982)

**757 Ando,T., Fowler,A.B., Stern,F.**
R Electronic properties of two-dimensional systems
Rev. Mod. Phys. 54    437-672 (1982)

**758 Ando,T., Mori,S.**
T Effective-mass theory of semiconductor heterojunctions and superlattices
Surf. Sci. 113    124-130 (1982)

**759 Ankri,D., Eastman,L.F.**
T GaAlAs-GaAs ballistic heterojunction bipolar transistor
Electron. Lett. 18    750-751 (1982)

**760 Aoki,H., Ando,T.**
T Effect of Landau-band structure on the quantized Hall conductivity in two dimensions
Surf. Sci. 113    27-31 (1982)

**761 Arakawa,Y., Sakaki,H.**
T Multidimensional quantum well laser and temperature dependence of its threshold current
Appl. Phys. Lett. 40    939-941 (1982)

**762 Asahi,H., Kawamura,Y., Nagai,H.**
Molecular beam epitaxial growth of InGaAlP on (100) GaAs
J. Appl. Phys. 53    4928-4931 (1982)

**763 Asahi,H., Kawamura,Y., Nagai,H., Ikegami,T.**
MBE growth of InGaAlP/InGaP/InGaAlP double heterostructures on (100) GaAs
Inst. Phys. Conf. Ser. 63    575-576 (1982)

**764 Asahi,H., Kawamura,Y., Nagai,H., Ikegami,T.**
Optically pumped laser action at 77 K of InGaP/InGaAlP double heterostructures grown by MBE
Electron. Lett. 18    62-63 (1982)

765 Asbeck,P.M., Miller,D.L., Asatourian,R., Kirkpatrick,C.G.
T  Numerical simulation of GaAs/GaAlAs heterojunction bipolar transistors
   IEEE Electron Device Lett. EDL-3    403-406 (1982)

766 Asbeck,P.M., Miller,D.L., Petersen,W.C., Kirkpatrick,C.G.
   GaAs/GaAlAs heterojunction bipolar transistors with cutoff frequencies above 10 GHz
   IEEE Electron Device Lett. EDL-3    366-368 (1982)

767 Bachrach,R.Z., Bringans,R.D.
   The Interaction of hydrogen with GaAs surfaces
   J. Physique 43 Colloque C5    C5/145-C5/151 (1982)

768 Bafleur,M., Munoz-Yague,A.
   Influence du processus d'elaboration sur les defauts cristallographiques dans les couches de GaAs epitaxiees par jets moleculaires
   J. Physique 43 Colloque C5    C5/465-C5/471 (1982)

769 Bafleur,M., Munoz-Yague,A., Rocher,A.
   Microtwinning and growth defects in GaAs MBE layers
   J. Cryst. Growth 59    531-538 (1982)

770 Ballingall,J.M., Wood,C.E.C.
   Crystal orientation dependence of silicon autocompensation in molecular beam epitaxial gallium arsenide
   Appl. Phys. Lett. 41    947-949 (1982)

771 Barnard,J.A., Wood,C.E.C., Eastman,L.F.
   Majority carrier light detectors with large gain-bandwidth products
   Inst. Phys. Conf. Ser. 63    461-466 (1982)

772 Barnard,J.A., Wood,C.E.C., Eastman,L.F.
   Preparation and properties of molecular beam epitaxy grown (Al(0.5)Ga(0.5)0.48)In(0.52)As
   IEEE Electron Device Lett. EDL-3    318-319 (1982)

773 Barrett,J.H.
T  Planar channeling of ions in compound semiconductor superlattices
   J. Vac. Sci. Technol. 21    384-385 (1982)

774 Barrett,J.H.
T  Mechanism of ion dechanneling in compound semiconductor superlattices
   Appl. Phys. Lett. 40    482-484 (1982)

775 Bastard,G.
T  Theoretical investigations of superlattice band structure in the envelope-function approximation
   Phys. Rev. B 25    7584-7596 (1982)

776  Bastard,G.
 T   Hydrogenic impurity states in a quantum well
     Surf. Sci. 113   165-169 (1982)

777  Bastard,G., Mendez,E.E., Chang,L.L., Esaki,L.
 T   Self-consistent calculations in InAs-GaSb
        heterojunctions
     J. Vac. Sci. Technol. 21   531-533 (1982)

778  Bastard,G., Mendez,E.E., Chang,L.L., Esaki,L.
 T   Exciton binding energy in quantum wells
     Phys. Rev. B 26   1974-1979 (1982)

779  Basu,B.K., Bhattacharya,K.
 T   Acoustic and piezoelectric scattering of
        two-dimensional electron gas in junction FET
        structures
     J. Phys. C 15   5711-5714 (1982)

780  Bauer,R.S.
 R   Interfaces of semiconducting molecular beam epitaxial
        films
     Thin Solid Films 89   419-432 (1982)

781  Bauer,R.S., Mikkelsen,J.C.,Jr.
     Surface processes controlling MBE heterojunction
        formation: GaAs(100)/Ge interfaces
     J. Vac. Sci. Technol. 21   491-497 (1982)

782  Beeby,J.L.
 T   Dynamical processes at surfaces
     Collected Papers of MBE-CST-2, Ed. R. Ueda (Jpn. Soc.
        Appl. Phys., Tokyo, 1982)   245-248 (1982)

783  Beneking,H., Cho,A.Y., Dekkers,J.J.M., Morkoc,H.
     Buried-channel GaAs MESFET's on MBE material:
        scattering parameters and intermodulation signal
        distortion
     IEEE Trans. Electron Devices ED-29   811-813 (1982)

784  Berezhkovskil,A.M., Ovchinnikov,A.A., Suris,R.A.
 T   Interband absorption in a semiconductor with a
        superlattice subjected to quantizing magnetic and
        electric fields
     Sov. Phys. Semicond. 16   1147-1150   *   Fiz. Tekh.
        Poluprovodn. 16   1788-1792 (1982)

785  Bethea,C.G., Chen,C.Y., Cho,A.Y., Garbinski,P.A.
     Opto-electronic picosecond sampling system utilizing
        a modulated barrier photodiode
     Appl. Phys. Lett. 40   591-594 (1982)

786 Bhattacharya,P.K., Buehlmann,H.J., Ilegems,M., Schmid,P., Melchior,H.
Properties of a Ga(x)In(1-x)As-GaAs isotype heterojunction diode
Appl. Phys. Lett 41   449-451 (1982)

787 Bhattacharya,P.K., Buehlmann,H.J., Ilegems,M., Staehli,J.L.
Impurity and defect levels in beryllium-doped GaAs grown by molecular beam epitaxy
J. Appl. Phys. 53   6391-6398 (1982)

788 Blanchet,R.C., Delhomme,B.J.
Correlation between deposition conditions, Auger analysis and electrical characteristics of MBE-grown SiO films on GaAs
Vacuum 32   3-8 (1982)

789 Blood,P., Harris,J.J., Joyce,B.A., Neave,J.H.
Deep states and surface processes in GaAs grown by molecular beam epitaxy
J. Physique 43 Colloque C5   C5/351-C5/355 (1982)

790 Blood,P., Roberts,J.S., Stagg,J.P.
GaInP grown by molecular beam epitaxy doped with Be and Sn
J. Appl. Phys. 53   3145-3149 (1982)

791 Bloss,W.L.
T   Optic and acoustic plasmon modes of a semiconductor superlattice
Solid State Commun. 44   363-367 (1982)

792 Bloss,W.L., Brody,E.M.
T   Collective modes of a superlattice - plasmons, lo phonon-plasmons, and magnetoplasmons
Solid State Commun. 43   523-528 (1982)

793 Bloss,W.L., Friedman,L.
T   Theory of optical mixing by mobile carriers in superlattices
Appl. Phys. Lett. 41   1023-1025 (1982)

794 Bluyssen,H.J.A., Maan,J.C., Wyder,P., Chang,L.L., Esaki,L.
Cyclotron resonance and Shubnikov-de Haas experiments in a n-InAs-GaSb superlattice
Phys. Rev. B 25   5364-5372 (1982)

795 Bonnevie,D., Huet,D.
Molecular beam epitaxial growth and characterization of In(0.53)Ga(0.47)As and InP substrate
J. Physique 43 Colloque C5   C5/445-C5/452 (1982)

796 **Briones,F., Collins,D.M.**
Low temperature photoluminescence of lightly Si-doped and undoped MBE GaAs
J. Electron. Mater. 11   847-866 (1982)

797 **Cadoret,R.**
R  Growth by vacuum evaporation, sputtering, molecular beam epitaxy and chemical vapor deposition
NATO Adv. Study Inst. Ser. C 87   453-488 (1982)

798 **Capasso,F.**
T  New ultra-low-noise avalanche photodiode with separated electron and hole avalanche regions
Electron. Lett. 18   12-13 (1982)

799 **Capasso,F.**
The channeling avalanche photodiode: a novel ultra-low-noise interdigitated p-n junction detector
IEEE Trans. Electron Devices ED-29   1388-1395 (1982)

800 **Capasso,F., Logan,R.A., Tsang,W.T.**
Interdigitated pn junction device with novel capacitance/voltage characteristic, ultralow capacitance and low punch-through voltage
Electron. Lett. 18   760-761 (1982)

801 **Capasso,F., Tsang,W.T., Hutchinson,A.L., Foy,P.W.**
The graded bandgap avalanche diode: a new molecular beam epitaxial structure with a large ionization rate ratio
Inst. Phys. Conf. Ser. 63   473 (1982)

802 **Capasso,F., Tsang,W.T., Hutchinson,A.L., Williams,G.F.**
Enhancement of electron impact ionization in superlattices: a new avalanche photodiode with a large ionization rate ratio
Inst. Phys. Conf. Ser. 63   569-570 (1982)

803 **Capasso,F., Tsang,W.T., Hutchinson,A.L., Williams,G.F.**
Enhancement of electron impact ionization in a superlattice: a new avalanche photodiode with a large ionization rate ratio
Appl. Phys. Lett. 40   38-40 (1982)

804 **Capasso,F., Tsang,W.T., Williams,G.F.**
R  New very low noise multilayer and graded-gap avalanche photodiodes for the 0.8 to 1.8 um wavelength region
Proc. SPIE - Inst. Soc. Opt. Eng. 340   50-55 (1982)

805 Carter,C.B., DeSimone,D., Griem,T., Wood,C.E.C.
Defects in heavily-doped MBE GaAs
Proc. 40th Annu. Meet. Electron Microsc. Soc. Am.
   40th    442-445 (1982)

806 Chai,Y.G., Chow,R.
Molecular beam epitaxial growth of lattice-mismatched
   In0.77Ga0.23As on InP
J. Appl. Phys. 53    1229-1232 (1982)

807 Chang,C.A.
Thermal removal of surface carbon from GaAs substrate
   used in molecular beam epitaxy
J. Vac. Sci. Technol. 21    663-665 (1982)

808 Chang,C.A.
Interface morphology studies of (110) and (111)
   Ge-GaAs grown by molecular beam epitaxy
Appl. Phys. Lett. 40    1037-1039 (1982)

809 Chang,C.A.
Interface morphology of epitaxial growth of Ge on
   GaAs and GaAs on Ge by molecular beam epitaxy
J. Appl. Phys. 53    1253-1255 (1982)

810 Chang,C.A.
 R  Interface studies of heterostructures
Collected Papers of MBE-CST-2, Ed. R. Ueda (Jpn. Soc.
   Appl. Phys., Tokyo, 1982)    131-134 (1982)

811 Chang,C.A., Takaoka,H., Chang,L.L., Esaki,L.
Molecular beam epitaxy of AlSb
Appl. Phys. Lett. 40    983-985 (1982)

812 Chang,L.L.
 R  Polytype heterostructures
Collected Papers of MBE-CST-2, Ed. R. Ueda (Jpn. Soc.
   Appl. Phys., Tokyo, 1982)    57-60 (1982)

813 Chang,L.L., Mendez,E.E., Kawai,N.J., Esaki,L.
Shubnikov-de Haas oscillations under tilted magnetic
   fields in InAs-GaSb superlattices
Surf. Sci. 113    306-312 (1982)

814 Chang,T.Y., Leheny,R.F., Nahory,R.E., Silberg,E.,
   Ballmann,A.A., Caridi,E.A., Harrold,C.J.
Junction field-effect transistors using
   In(0.53)Ga(0.47)As material grown by molecular
   beam epitaxy
IEEE Electron Device Lett. EDL-3    56-58 (1982)

815 Chang,Y.C., Schulman,J.N.
 T  Theory of heterostructures: A reduced Hamiltonian
      method with evanescent states and transfer matrices
J. Vac. Sci. Technol. 21    540-543 (1982)

816  Chen,C.Y., Bethea,C.G., Cho,A.Y., Garbinski,P.A.
Temporal resolution of an Al(x)Ga(1-x)As/GaAs
    bias-free photodetector
Electron. Lett. 18    890-891 (1982)

817  Chen,C.Y., Cho,A.Y., Alavi,K., Garbinski,P.A.
Short Channel Ga(0.47)In(0.53)As/Al(0.48)In(0.52)As
    selectively doped field effect transistors
IEEE Electron Device Lett. EDL-3    205-208 (1982)

818  Chen,C.Y., Cho,A.Y., Bethea,C.G., Garbinski,P.A.
Bias-free selectively doped Al(x)Ga(1-x)As-GaAs
    picosecond photodetectors
Appl. Phys. Lett. 41    282-284 (1982)

819  Chen,C.Y., Cho,A.Y., Cheng,K.Y., Garbinski,P.A.
Quasi-Schottky barrier diode on n-Ga0.47In0.53As
    using a fully depleted p+ -Ga0.47In0.53As layer
    grown by molecular beam epitaxy
Appl. Phys. Lett. 40    401-403 (1982)

820  Chen,C.Y., Cho,A.Y., Cheng,K.Y., Pearsall,T.P.,
    O'Connor,P., Garbinski,P.A.
Depletion mode modulation doped
    Al0.48In0.52As-Ga0.47In0.53As heterojunction field
    effect transistors
IEEE Electron Device Lett. EDL-3    152-155 (1982)

821  Chen,C.Y., Cho,A.Y., Garbinski,P.A., Bethea,C.G.
GaAs/AlxGa1-xAs depletion stop phototransistor grown
    by molecular beam epitaxy
Appl. Phys. Lett. 40    510-512 (1982)

822  Chen,C.Y., Cho,A.Y., Garbinski,P.A., Cheng,K.Y.
Characteristics of an In(0.53)Ga(0.47)As very shallow
    junction gate structure grown by molecular beam
    epitaxy
IEEE Electron Device Lett. EDL-3    15-17 (1982)

823  Chen,C.Y., Cho,A.Y., Gossard,A.C., Garbinski,P.A.
Offset channel insulated gate field-effect transistors
Appl. Phys. Lett. 41    360-362 (1982)

824  Chen,P., Bolmont,D., Sebenne,C.A.
Preparation and electronic properties of abrupt
    Ge-GaAs(110) interfaces
J. Phys. C 15    6101-6111 (1982)

825  Cheng,K.Y., Cho,A.Y.
Silicon doping and impurity profiles in
    Ga0.47In0.53As and Al0.48In0.52As grown by
    molecular beam epitaxy
J. Appl. Phys. 53    4411-4415 (1982)

826 **Cheng,K.Y., Cho,A.Y.**
Electron mobilities in modulation doped
Ga0.47In0.53As/Al0.48In0.52As heterojunctions
grown by molecular beam epitaxy
Appl. Phys. Lett. 40    147-149 (1982)

827 **Cheng,K.Y., Cho,A.Y.**
The growth of high purity Ga(0.47)In(0.53)As on InP
by MBE
Collected Papers of MBE-CST-2, Ed. R. Ueda (Jpn. Soc.
Appl. Phys., Tokyo, 1982)    103-106 (1982)

828 **Cheng,K.Y., Cho,A.Y., Christman,S.B.,
Pearsall,T.P., Rowe,J.E.**
Measurement of the gamma-L separation in
Ga0.47In0.53As by ultraviolet photoemission
Appl. Phys. Lett. 40    423-425 (1982)

829 **Chiang,T.C., Ludeke,R., Eastman,D.E.**
Photoemission studies of Al(x)Ga(1-x)As(100) surfaces
grown by molecular-beam epitaxy
Phys. Rev. B 25    6518-6521 (1982)

830 **Cho,A.Y.**
R  Overview on molecular beam epitaxy
Collected Papers of MBE-CST-2, Ed. R. Ueda (Jpn. Soc.
Appl. Phys., Tokyo, 1982)    3-7 (1982)

831 **Cho,A.Y., Kollberg,E., Zirath,H., Snell,W.W.,
Schneider,M.V.**
Single-crystal metal-semiconductor microjunctions
prepared by molecular beam epitaxy
Electron. Lett. 18    424-425 (1982)

832 **Chomette,A., Palmier,J.F.**
T  Tenseur de mobilite d'un super-reseau en presence de
desordre d'interface
Solid State Commun. 43    157-161 (1982)

833 **Christou,A., Davey,J.E., Covington,D.W.**
Thin MBE GaAs millimetre-wave mixer diode using Ge
substrate
Electron. Lett. 18    367-368 (1982)

834 **Chu,W.K., Saris,F.W., Chang,C.A., Ludeke,R.,
Esaki,L.**
Ion-beam crystallography of InAs-GaSb superlattices
Phys. Rev. B 26    1999-2010 (1982)

835 **Ciccacci,F., Alvarado,S.F., Valeri,S.**
Spin-polarized photoelectron emission study of
$Al_xGa_{1-x}As$ alloys grown by molecular beam epitaxy
J. Appl. Phys. 53    4395-4398 (1982)

836  Clegg,J.B., Foxon,C.T., Weimann,G.
     Secondary ion mass spectrometry study of lightly
        doped p-type GaAs films grown by molecular beam
        epitaxy
     J. Appl. Phys. 53    4518-4520 (1982)

837  Coleman,P.D.
     Real space transfer electron device oscillator - a
        new candidate for the near millimeter range
     Proc. SPIE - Int. Soc. Opt. Eng. 317    333-338 (1982)

838  Coleman,P.D., Freeman,J., Morkoc,H., Hess,K.,
        Streetman,B.G., Keever,M.
     Demonstration of a new oscillator based on real-space
        transfer in heterojunctions
     Appl. Phys. Lett. 40    493-495 (1982)

839  Collins,D.M.
     On the use of "downward-looking" sources in MBE
        systems
     J. Vac. Sci. Technol. 20    250-251 (1982)

840  Collins,D.M., Miller,N.J.
     Sn and Te doping of molecular beam epitaxial GaAs
        using a SnTe source
     J. Appl. Phys. 53    3010-3018 (1982)

841  Cooper,J.A.,Jr., Capasso,F., Thornber,K.K.
  T  Semiconductor structures for repeated velocity
        overshoot
     IEEE Electron Device Lett. EDL-3    407-408 (1982)

842  Cox,N.W.
  R  Application of molecular beam epitaxy to microwave
        and millimeter wave devices
     Proc. SPIE - Int. Soc. Opt. Eng. 317    325-332 (1982)

843  Cullis,A.G., Farrow,R.F.C., Chew,N.G.,
        Williams,G.M.
     Electron microscope studies of the semiconductor to
        metal phase changes in epitaxial layers of
        elemental tin
     Inst. Phys. Conf. Ser. 61    535-538 (1982)

844  Daeweritz,L.
  T  Facetting, steps and reconstruction on GaAs(001)
     Surf. Sci. 118    585-596 (1982)

845  Das Sarma,S.
  T  Coupled electron-LO phonon modes of
        GaAs-Al(x)Ga(1-x)As multilayer systems
     Appl. Surf. Sci. 11/12    535-543 (1982)

846  Das Sarma,S., Quinn,J.J.
  T  Collective excitations in semiconductor superlattices
     Phys. Rev. B 25    7603-7618 (1982)

847 DeSimone,D., Wood,C.E.C., Evans,C.A.,Jr.
Manganese incorporation behavior in molecular beam epitaxial gallium arsenide
J. Appl. Phys. 53   4938-4942 (1982)

848 Delagebeaudeuf,D., Delescluse,P., Etienne,P., Massies,J., Laviron,M., Chaplart,J., Linh,N.T.
Tunneling through GaAs-Al(x)Ga(1-x)As-GaAs double heterojunctions
Electron. Lett. 18   85-87 (1982)

849 Delagebeaudeuf,D., Laviron,M., Delescluse,P., Tung,P.N., Chaplart,J., Linh,N.T.
Planar enhancement mode two-dimensional electron gas FET associated with a low AlGaAs surface potential
Electron. Lett. 18   103-105 (1982)

850 Delagebeaudeuf,D., Linh,N.T.
T  Metal-(n)AlGaAs-GaAs two-dimensional electron gas FET
IEEE Trans. Electron Devices ED-29   955-960 (1982)

851 Delagebeaudeuf,D., Linh,N.T.
T  Speed power in planar two-dimensional electron gas FET DCFL circuit: a theoretical approach
Electron. Lett. 18   510-512 (1982)

852 Delescluse,P., Delagebeaudeuf,D., Tung,P.N., Rochette,J.F., Laviron,M., Etienne,P., Massies,J., Linh,N.T.
MBE growth of GaAs/n-AlGaAs for FET applications
Collected Papers of MBE-CST-2, Ed. R. Ueda (Jpn. Soc. Appl. Phys., Tokyo, 1982)   117-120 (1982)

853 DiLorenzo,J.V., Dingle,R., Feuer,M., Gossard,A.C., Hendel,R., Hwang,J.C.M., Kastalsky,A., Keramidas,V.G., Kiehl,R.A., O'Connor,P.
R  Material and device considerations for selectively doped heterojunction transistors
IEDM 82 Technical Digest   578-581 (1982)

854 Dingle,R., Weisbuch,C., Stoermer,H.L., Morkoc,H., Cho,A.Y.
Characterization of high purity GaAs grown by molecular beam epitaxy
Appl. Phys. Lett. 40   507-510 (1982)

855 Dobson,P.J., Neave,J.H., Joyce,B.A.
RHEED evidence for a domain structure of GaAs(001)-2x4 and -4x2 reconstructed surfaces
Surf. Sci. 119   L339-L345 (1982)

856 Dobson,P.J., Scott,G.B., Neave,J.H., Joyce,B.A.
The occurence of sharp exciton-like features in low temperature photoluminescence spectra from MBE grown GaAs
Solid State Commun. 43   917-919 (1982)

857 Doehler,G.H.
T   Semiconductors with n-i-p-i doping superlattices -
    electrooptical device aspects
    Collected Papers of MBE-CST-2, Ed. R. Ueda (Jpn. Soc.
       Appl. Phys., Tokyo, 1982)   21-24 (1982)

858 Doehler,G.H., Kuenzel,H., Ploog,K.
    Tunable absorption coefficient in GaAs doping
       superlattices
    Phys. Rev. B 25   2616-2626 (1982)

859 Dordzhin,G.S., Sadof'ev,Y.G., Senichkina,R.S.,
       Sharonova,L.V., Shik,A.Y.
    Silicon-gallium arsenide heterojunctions prepared by
       the method of molecular beam epitaxy
    Sov. Phys. Semicond. 16   1057-1058   *   Fiz. Tekh.
       Poluprovodn. 16   1654-1656 (1982)

860 Drummond,T.J., Fischer,R., Miller,P.A., Morkoc,H.,
       Cho,A.Y.
    Influence of substrate temperature on electron
       mobility in normal and inverted single period
       modulation doped Al(x)Ga(1-x)As/GaAs
       heterojunctions
    J. Vac. Sci. Technol. 21   684-688 (1982)

861 Drummond,T.J., Fischer,R., Morkoc,H., Miller,P.A.
    Influence of substrate temperature on the mobility of
       modulation-doped AlxGa1-xAs/GaAs heterostructures
       grown by molecular beam epitaxy
    Appl. Phys. Lett. 40   430-432 (1982)

862 Drummond,T.J., Keever,M., Morkoc,H.
    Comparison of single and multiple period modulation
       doped Al(x)Ga(1-x)As/GaAs heterostuctures for FETs
    Jpn. J. Appl. Phys. 21   L65-L67 (1982)

863 Drummond,T.J., Kopp,W., Fischer,R., Morkoc,H.
    Influence of AlAs mole fraction on the electron
       mobility of (Al,Ga)As/GaAs heterostructures
    J. Appl. Phys. 53   1028-1029 (1982)

864 Drummond,T.J., Kopp,W., Fischer,R., Morkoc,H.,
       Thorne,R.E., Cho,A.Y.
    Photoconductivity effects in extremely high mobility
       modulation-doped (Al,Ga)As
    J. Appl. Phys. 53   1238-1240 (1982)

865 Drummond,T.J., Kopp,W., Keever,M., Morkoc,H.,
       Cho,A.Y.
    Electron mobility in single and multiple period
       modulation-doped (Al,Ga)As/GaAs heterostructures
    J. Appl. Phys. 53   1023-1027 (1982)

866   Drummond,T.J., Kopp,W., Morkoc,H., Keever,M.
      Transport in modulation-doped structures
         (Al(x)Ga(1-x)As/GaAs) and correlations with Monte
         Carlo calculations (GaAs)
      Appl. Phys. Lett. 41    277-279 (1982)

867   Drummond,T.J., Kopp,W., Thorne,R.E., Fischer,R.,
         Morkoc,H.
      Influence of AlxGa1-xAs buffer layers on the
         performance of modulation-doped field-effect
         transistors
      Appl. Phys. Lett. 40    879-881 (1982)

868   Drummond,T.J., Lyons,W.G., Fischer,R., Thorne,R.E.,
         Morkoc,H., Hopkins,C.G., Evans,C.A.,Jr.
      Si incorporation in As(x)Ga(1-x)As grown by molecular
         beam epitaxy
      J. Vac. Sci. Technol. 21    957-960 (1982)

869   Drummond,T.J., Morkoc,H., Cheng,K.Y., Cho,A.Y.
      Current transport in modulation-doped
         Ga0.47In0.53As/Al0.48In0.52As heterojunctions at
         moderate electric fields
      J. Appl. Phys. 53    3654-3657 (1982)

870   Drummond,T.J., Morkoc,H., Cho,A.Y.
 R    Molecular beam epitaxy growth of (Al,Ga)As/GaAs
         heterostructures
      J. Cryst. Growth 56    449-454 (1982)

871   Drummond,T.J., Su,S.L., Kopp,W., Fischer,R.,
         Thorne,R.E., Morkoc,H., Lee,K., Shur,M.S.
      High-velocity n-on and n-off modulation doped
         GaAs/Al(x)Ga(1-x)As FETs
      IEDM 82 Technical Digest    586-589 (1982)

872   Drummond,T.J., Su,S.L., Lyons,W.G., Fischer,R.,
         Kopp,W., Morkoc,H., Lee,K., Shur,M.S.
      Enhancement of electron velocity in modulation-doped
         (Al,Ga)As/GaAs FETs at cryogenic temperatures
      Electron. Lett. 18    1057-1058 (1982)

873   Drummond,T.J., Wang,T., Kopp,W., Morkoc,H.,
         Thorne,R.E., Su,S.L.
      A novel normally-off camel diode gate GaAs
         field-effect transistor
      Appl. Phys. Lett. 40    834-836 (1982)

874   Duggan,G., Dawson,P., Foxon,C.T., Hooft,G.W. t'
      The effect of arsenic species on the minority carrier
         properties of (AlGa)As-GaAs double
         heterostructures grown by MBE
      J. Physique 43 Colloque C5    C5/129-C5/134 (1982)

875　Eastman,L.F.
　　　The limits of electron ballistic motion in compound
　　　　　semiconductor transistors
　　　Inst. Phys. Conf. Ser. 63　　245-250　(1982)

876　Eastman,L.F.
　R　The limits of ballistic motion in compound
　　　　　semiconductor transistors
　　　Inst. Phys. Conf. Ser. 63　　245-250　(1982)

877　Eastman,L.F.
　R　Very high electron velocity in short gallium arsenide
　　　　　structures
　　　in "Festkoerperprobleme", Ed. J. Treusch (Vieweg,
　　　　　Braunschweig, 1982) Vol. XXII　　173-187　(1982)

878　Ebert,G., Klitzing,K. von, Probst,C., Ploog,K.
　　　Magneto-quantumtransport on GaAs-Al(x)Ga(1-x)As
　　　　　heterostructures at very low temperatures
　　　Solid State Commun. 44　　95-98　(1982)

879　Englert,T., Tsui,D.C., Gossard,A.C., Uihlein,C.
　　　g-factor enhancement in the 2D electron gas in
　　　　　GaAs/AlGaAs heterojunctions
　　　Surf. Sci. 113　　295-300　(1982)

880　Englert,T., Tsui,D.C., Portal,J.C., Beerens,J.,
　　　　　Gossard,A.C.
　　　Magnetophonon resonances of the two-dimensional
　　　　　electron gas in GaAs/AlGaAs heterostructures
　　　Solid State Commun. 44　　1301-1304　(1982)

881　Esaki,L.
　R　Advances in synthesized superlattices
　　　Lect. Notes Phys. 152　　340-351　(1982)

882　Esaki,L.
　R　Advances in synthesized superlattices
　　　in "Novel Materials and Techniques in Condensed
　　　　　Matter", Eds. G.W. Crabtree and P. Vashishta
　　　　　(Elsevier, New York, 1982)　　1-19　(1982)

883　Farrow,R.F.C., Sullivan,P.W., Williams,G.M.,
　　　　　Stanley,C.R.
　　　The performance of a double oven as an arsenic dimer
　　　　　source for MBE
　　　Collected Papers of MBE-CST-2, Ed. R. Ueda (Jpn. Soc.
　　　　　Appl. Phys., Tokyo, 1982)　　169-172　(1982)

884　Favennec,P.N., Henry,L., Regreny,A., Salvi,M.
　　　Selected-area molecular beam epitaxy on ion-implanted
　　　　　GaAs substrates
　　　Electron. Lett. 18　　933-935　(1982)

885 Feng,M., Eu,V.K., D'Haenens,I.J., Braunstein,M.
Low-noise GaAs fiel-effect transistor made by molecular beam epitaxy
Appl. Phys. Lett. 41   633-635 (1982)

886 Fischer,A., Graf,K., Hafendoerfer,M., Kuenzel,H., Ploog,K.
Rechnersteuerung fuer den Betrieb von Molekularstrahl-Epitaxie-Anlagen: 1) Grundlagen und Aufbau der Hardware   2) Programmsystem und Anwendung bei der Epitaxie
Techn. Messen 49   403-408, 461-469 (1982)

887 Foxon,C.T., Dawson,P., Duggan,G., Hooft,G.W. t'
The effect of arsenic species, As2 or As4, on the optical properties of (AlGa)As-GaAs double heterostructures grown by MBE
Collected Papers of MBE-CST-2, Ed. R. Ueda (Jpn. Soc. Appl. Phys., Tokyo, 1982)   81-84 (1982)

888 Freller,H., Guenther,K.G.
R   Three-temperature method as an origin of molecular beam epitaxy
Thin Solid Films 88   291-307 (1982)

889 Fujii,T., Hiyamizu,S., Wada,O., Sugahara,T., Yamakoshi,S., Ishikawa,T., Sakurai,T., Hashimoto,H.
Extremely uniform GaAs-AlGaAs heterostructure layers with high optical quality by MBE
Collected Papers of MBE-CST-2, Ed. R. Ueda (Jpn. Soc. Appl. Phys., Tokyo, 1982)   85-88 (1982)

890 Fukuyama,H., Platzman,P.M.
T   Coulomb correlations and the quantum Hall effect
Phys. Rev. B 25   2934-2936 (1982)

891 Gant,H., Moench,W.
On the chemisorption of Ge on GaAs(100) surfaces: UPS and work function measurements
Appl. Surf. Sci. 11/12   332-347 (1982)

892 Ghibaudo,G., Kamarinos,G.
T   Thermopower of a quasi-two-dimensional electron gas
Phys. Status Solidi B 114   K105-K110 (1982)

893 Gibbs,H.M., Jewell,J.L., Moloney,J.V., Tarng,S.S., Tai,K., Watson,E.A., Gossard,A.C., McCall,S.L., Passner,A., Venkatesan,T.N.C. ,Wiegmann,W.
Switching of a GaAs bistable etalon: external switching on and off, regenerative pulsations, transverse effects, and lasing
Proc. SPIE - Int. Soc. Opt. Eng. 321   67-74 (1982)

894 Gibbs,H.M., Tarng,S.S., Jewell,J.L.,
      Weinberger,D.A., Tai,K., Gossard,A.C.,
      McCall,S.L., Passner,A., Wiegmann,W.
    Room-temperature excitonic optical bistability in a
      GaAs-GaAlAs superlattice etalon
    Appl. Phys. Lett. 41   221-222 (1982)

895 Goldstein,L., Quillec,M., Rao,E.V.K., Henoc,P.,
      Masson,J.M., Marzin,J.Y.
    Preparation and characterization of strained
      superlattices structures of InGaAs/GaAs by MBE
    J. Physique 43 Colloque C5   C5/201-C5/207 (1982)

896 Goncalves da Silva,C.E.T., Fulco,P.
  T  Electronic states associated with extended defects in
      quantum well structures
    Rev. Bras. Fis. 12   325-336 (1982)

897 Gossard,A.C.
  R  New phenomena in superlattice structures
    Collected Papers of MBE-CST-2, Ed. R. Ueda (Jpn. Soc.
      Appl. Phys., Tokyo, 1982)   35-38 (1982)

898 Gossard,A.C.
  R  Molecular beam epitaxy of superlattices in thin films
    in "Treatise on Material Science and Technology" Vol.
      24: 'Thin Films: Preparation and Properties',Eds.
      K.N.Tu and R.Rosenberg (Academic Press,New
      York,1982)   13-66 (1982)

899 Gossard,A.C., Brown,W., Allyn,C.L., Wiegmann,W.
    Molecular beam epitaxial growth and electrical
      transport of graded barriers for nonlinear current
      conduction
    J. Vac. Sci. Technol. 20   694-700 (1982)

900 Gossard,A.C., Kazarinov,R.F., Luryi,S.,
      Wiegmann,W.
    Electric properties of unipolar GaAs structures with
      ultrathin triangular barriers
    Appl. Phys. Lett. 40   832-833 (1982)

901 Gossard,A.C., Wiegmann,W., Miller,R.C.,
      Petroff,P.M., Tsang,W.T.
    Growth of single-quantum-well structures by molecular
      beam epitaxy
    Collected Papers of MBE-CST-2, Ed. R. Ueda (Jpn. Soc.
      Appl. Phys., Tokyo, 1982)   39-42 (1982)

902 Gotoh,H., Yamamoto,T., Kimata,M.
    Molecular beam epitaxy of In(0.2)Ga(0.8)Sb
    Jpn. J. Appl. Phys. 21   L767-L769 (1982)

903 Grange,J.D.
R   System related impurity incorporation during the
        growth of gallium arsenide by molecular beam
        epitaxy
    Vacuum 32    477-480 (1982)

904 Grondin,R.O., Lugli,P., Ferry,D.K.
T   Ballistic transport in semiconductors
    IEEE Electron Device Lett. EDL-3    373-375 (1982)

905 Guldner,Y., Vieren,J.P., Voisin,P., Voos,M.,
    Maan,J.C., Chang,L.L., Esaki,L.
    Observation of double cyclotron resonance and
        interband transitions in InAs-GaSb
        multi-heterojunctions
    Solid State Commun. 41    755-758 (1982)

906 Halperin,B.I.
T   Quantized Hall conductance, current-carrying edge
        states, and the existence of extended states in a
        two-dimensional disordered potential
    Phys. Rev. B 25    2185-2190 (1982)

907 Harris,J.H., Sugai,S., Nurmikko,A.V.
    Interface recombination and carrier confinement at a
        GaAs/Ga(x)In(1-x)P double heterojunction studied
        by picosecond population modulation spectroscopy
    Appl. Phys. Lett. 40    885-887 (1982)

908 Harris,J.J., Joyce,B.A., Gowers,J.P., Neave,J.H.
    Nucleation effects during MBE growth of Sn-doped GaAs
    Appl. Phys. A 28    63-71 (1982)

909 Hegarty,J., Sturge,M.D., Gossard,A.C., Wiegmann,W.
    Resonant degenerate four-wave mixing in GaAs
        multiquantum well structures
    Appl. Phys. Lett. 40    132-134 (1982)

910 Hegarty,J., Sturge,M.D., Weisbuch,C., Gossard,A.C.,
    Wiegmann,W.
    Resonant Rayleigh scattering from an inhomogeneously
        broadened transition: a new probe of the
        homogeneous linewidth
    Phys. Rev. Lett. 49    930-932 (1982)

911 Heiblum,M., Nathan,M.I., Chang,C.A.
    Charactersitics of AuGeNi ohmic contacts to GaAs
    Solid-State Electron. 25    185-195 (1982)

912 Herman,M.A.
R   Physical problems concerning effusion processes of
        semiconductors in molecular beam epitaxy
    Vacuum 32    555-565 (1982)

913   Hierl,T.L., Luscher,P.E.
      Production of microwave devices by MBE
      Collected Papers of MBE-CST-2, Ed. R. Ueda (Jpn. Soc.
         Appl. Phys., Tokyo, 1982)    147-150 (1982)

914   Hikosaka,K., Mimura,T., Hiyamizu,S.
      Deep electron traps in MBE-grown AlGaAs ternary alloy
         for heterojunction devices
      Inst. Phys. Conf. Ser. 63    233-238 (1982)

915   Hiyamizu,S.
  R   Recent developments in MBE GaAs/n-AlGaAs
         heterostructures and HEMTs
      Collected Papers of MBE-CST-2, Ed. R. Ueda (Jpn. Soc.
         Appl. Phys., Tokyo, 1982)    113-116 (1982)

916   Hiyamizu,S., Mimura,T.
  R   High mobility electrons in selectively doped
         GaAs/n-AlGaAs heterostructures grown by MBE and
         their application to high-speed devices
      J. Cryst. Growth 56    455-463 (1982)

917   Hiyamizu,S., Mimura,T.
  R   MBE-grown selectiveliy doped GaAs/n-AlGaAs
         heterostructures and their application to high
         electron mobility transistors
      Jpn. Annu. Rev. Electron., Comput. & Telecommun.:
         Semicond. Technol. 7    258-271 (1982)

918   Hiyamizu,S., Mimura,T., Ishikawa,T.
  R   MBE-grown GaAs/N-AlGaAs heterostructures and their
         application to high electron mobility transistors
      Jpn. J. Appl. Phys. 21, Suppl. 21-1    161-168 (1982)

919   Hjalmarson,H.P.
  T   Band structure of impurity-sheet-doped superlattice
         alloys
      J. Vac. Sci. Technol. 21    524-527 (1982)

920   Hoepfel,R.A., Lindemann,G., Gornik,E., Stangl,G.,
         Gossard,A.C., Wiegmann,W.
      Cyclotron and plasmon emission from two-dimensional
         electrons in GaAs
      Surf. Sci. 113    118-123 (1982)

921   Holah,G.D., Eisele,F.L., Meeks,E.L., Cox,N.W.
      Growth of InGaAsP by molecular beam epitaxy
      Appl. Phys. Lett. 41    1073-1075 (1982)

922   Holonyak,N.,Jr., Vojak,B.A., Morkoc,H.,
         Drummond,T.J., Hess,K.
      Stimulated emission in a degenerately doped GaAs
         quantum well
      Appl. Phys. Lett. 40    658-660 (1982)

923 Holonyak,N.,Jr., Vojak,B.A., Morkoc,H.,
    Drummond,T.J., Hess,K.
    Stimulated emission in a degenerately doped GaAs
        quantum well
    Appl. Phys. Lett. 40    658-660 (1982)

924 Hoskins,M.J., Morkoc,H., Hunsinger,B.J.
    Charge transport by surface acoustic waves in GaAs
    Appl. Phys. Lett. 41    332-334 (1982)

925 Hotta,T., Sakaki,H., Ohno,H.
    A new AlGaAs/GaAs heterojunction FET with insulated
        gate structure (MISSFET)
    Jpn. J. Appl. Phys. 21    L122-L124 (1982)

926 Hove,J.M. van, Cohen,P.I.
    Development of steps on GaAs during molecular beam
        epitaxy
    J. Vac. Sci. Technol. 20    726-729 (1982)

927 Hwang,J.C.M., DiLorenzo,J.V., Luscher,P.E.,
    Knodle,W.S.
    Application of molecular beam epitaxy to III-V
        microwave and high speed device fabrication
    Solid State Technol. 25 No. 10    166-169 (1982)

928 Hwang,J.C.M., Flahive,P.G., Wemple,S.H.
    Performance of power FET's fabricated on MBE-grown
        GaAs layers
    IEEE Electron Device Lett. EDL-3    320-321 (1982)

929 Iafrate,G.J., Ferry,D.K., Reich,R.K.
T   Lateral (two-dimensional) superlattices: quantum-well
        confinement and charge instabilities
    Surf. Sci. 113    485-488 (1982)

930 Imry,Y.
T   The quantized Hall effect and other macroscopic
        quantum phenomena
    in "Anderson Localization", Eds. Y. Nagaoka and H.
        Fukuyama, Springer Ser. Solid-State Sci. Vol. 39
        198-206 (1982)

931 Inoue,M., Hiyamizu,S., Hida,H., Nanbu,K.,
    Hashimoto,H., Inuishi,Y.
    Transport properties of 2D hot electrons at
        modulation-doped GaAs/AlGaAs interfaces
    Inst. Phys. Conf. Ser. 63    257-262 (1982)

932 Iordansky,S.V.
T   On the conductivity of two dimensional electrons in a
        strong magnetic field
    Solid State Comm. 43    1-3 (1982)

933   Ishibashi,T., Tarucha,S., Okamoto,H.
      Si and Sn doping in AL(x)Ga(1-x)As grown by MBE
      Jpn. J. Appl. Phys. 21   L476-L478 (1982)

934   Ishibashi,T., Tarucha,S., Okamoto,H.
      Si and Sn doping in Al(x)Ga(1-x)As grown by MBE
      Collected Papers of MBE-CST-2, Ed. R. Ueda (Jpn. Soc.
         Appl. Phys., Tokyo, 1982)   25-28 (1982)

935   Ishibashi,T., Tarucha,S., Okamoto,H.
      Exciton associated optical absorption spectra of
         AlAs/GaAs superlattices at 300 K
      Inst. Phys. Conf. Ser. 63   587-588 (1982)

936   Ishikawa,T., Saito,J., Sasa,S., Hiyamizu,S.
      Electrical properties of Si-doped Al(x)Ga(1-x)As
         layers grown by MBE
      Jpn. J. Appl. Phys. 21   L675-L676 (1982)

937   Iwamura,H., Saku,T., Ishibashi,T., Naganuma,M.,
         Okamoto,H.
      Comparison between a conventional DH and an MQW laser
         grown by MBE
      Collected Papers of MBE-CST-2, Ed. R. Ueda (Jpn. Soc.
         Appl. Phys., Tokyo, 1982)   47-50 (1982)

938   Jenkinson,H.A., Zavada,J.M., Laidig,W.D.,
         Wortman,J.J., Littlejohn,M.A.
      GaAs MBE layers for infrared optical waveguiding
      Collected Papers of MBE-CST-2, Ed. R. Ueda (Jpn. Soc.
         Appl. Phys., Tokyo, 1982)   95-98 (1982)

939   Jewell,J.L., Gibbs,H.M., Tarng,S.S., Gossard,A.C.,
         Wiegmann,W.
      Regenerative pulsations from an intrinsic bistable
         optical device
      Appl. Phys. Lett. 40   291-293 (1982)

940   Jiang,D.S., Makita,Y., Ploog,K., Queisser,H.J.
      Electrical properties and photoluminescence of
         Te-doped GaAs grown by molecular beam epitaxy
      J. Appl. Phys. 53   999-1006 (1982)

941   Joyce,B.A.
  R   Surface processes and film properties in MBE growth
         of III-V compounds
      Collected Papers of MBE-CST-2, Ed. R. Ueda (Jpn. Soc.
         Appl. Phys., Tokyo, 1982)   9-13 (1982)

942   Joyce,B.A.
  R   Kinetics, mechanisms and surface processes in
         molecular beam epitaxy of III-V compounds and
         alloys
      in "Proc. Int. Workshop Phys. Semicond. Devices",
         Eds. S.C. Jain and S. Radhakrishna (Wiley East,
         New Delhi, 1982)   542-550 (1982)

943 Jung,H., Doehler,G.H., Kuenzel,H., Ploog,K., Ruden,P., Stolz,H.J.
Photoluminescence study of electron-hole recombination across the tunable effective gap in GaAs n-i-p-i superlattices
Solid State Commun. 43   291-294 (1982)

944 Jung,H., Kuenzel,H., Ploog,K.
Influence of arsenic vapor species on electrical and optical properties of MBE grown GaAs
J. Physique 43 Colloque C5   C5/135-C5/143 (1982)

945 Kahn,A., Carelli,J., Miller,D.L., Kowalczyk,S.P.
Comparative LEED studies of Al(x)Ga(1-x)As(110) and GaAs(110)-Al
J. Vac. Sci. Technol. 21   380-383 (1982)

946 Kastalsky,A., Dingle,R., Cheng,K.Y., Cho,A.Y.
Two-dimensional electron gas at molecular beam epitaxial-grown, selectively doped, In(0.53)Ga(0.47)As-In(0.48)Al(0.52)As interface
Appl. Phys. Lett. 41   274-277 (1982)

947 Kawabe,M., Matsuura,N., Inuzuka,H.
Composition control of Al(x)Ga(1-x)As and new type of superlattice by pulsed molecular beam
Jpn. J. Appl. Phys. 21   L447-L448 (1982)

948 Kawabe,M., Matsuura,N., Toda,K., Inuzuka,H.
A new composition control method of Al(x)Ga(1-x)As by molecular beam epitaxy
Jpn. J. Appl. Phys. 21, Suppl. 21-1   439-440 (1982)

949 Kawai,N.J., Wood,C.E.C., Eastman,L.F.
Carrier compensation at interfaces formed by molecular beam epitaxy
J. Appl. Phys. 53   6208-6213 (1982)

950 Kawamura,Y., Asahi,H., Nagai,H., Ikegami,T.
Doping studies of InGaP and InGaAlP by MBE
Collected Papers of MBE-CST-2, Ed. R. Ueda (Jpn. Soc. Appl. Phys., Tokyo, 1982)   99-102 (1982)

951 Kawamura,Y., Noguchi,Y., Asahi,H., Nagai,H.
MBE-grown InGaAs/InP BH lasers with LPE burying layers
Electron. Lett. 18   91-92 (1982)

952 Kawanami,H., Sakamoto,T., Takahashi,T., Suzuki,E., Nagai,K.
Heteroepitaxial growth of GaP on a Si(100) substrate by molecular beam epitaxy
Jpn. J. Appl. Phys. 21   L68-L70 (1982)

953 Kazarinov,R.F., Luryi,S.
T   Quantum percolation and quantization of Hall
        resistance in two-dimensional electron gas
    Phys. Rev. B 25   7626-7630 (1982)

954 Keever,M., Drummond,T.J., Morkoc,H., Hess,K.,
        Streetman,B.G., Ludowise,M.
    Hall effect and mobility in heterojunction layers
    J. Appl. Phys. 53   1034-1036 (1982)

955 Keever,M., Kopp,W., Drummond,T.J., Morkoc,H.,
        Hess,K.
    Current transport in modulation-doped
        Al(x)Ga(1-x)As/GaAs heterojunction structures at
        moderate field strengths
    Jpn. J. Appl. Phys. 21   1489-1495 (1982)

956 Kido,G., Miura,N., Ohno,H., Sakaki,H.
    Magnetophonon resonance in a two-dimensional electron
        system in the GaAs-Al(x)Ga(1-x)As heterojunction
        interface
    J. Phys. Soc. Jpn. 51   2168-2173 (1982)

957 Kim,J.Y., Madhukar,A.
T   Electronic structure of GaP-AlP(100) superlattices
    J. Vac. Sci. Technol. 21   528-530 (1982)

958 Kitahara,K., Nakai,K., Shibatomi,A., Ohkawa,S.
    Current limitation induced by infrared light in
        n-type GaAs thin layers on semi-insulating
        Cr-doped GaAs
    Jpn. J. Appl. Phys. 21   513-516 (1982)

959 Klitzing,K. von
R   Two-dimensional systems: A method for the
        determination of the fine structure constant
    Surf. Sci. 113   1-9 (1982)

960 Kondo,K., Muto,S., Nanbu,K., Ishikawa,T.,
        Hiyamizu,S., Hashimoto,H.
    Effect of H2 on the quality of Si-doped
        Al(x)Ga(1-x)As grown by MBE
    Collected Papers of MBE-CST-2, Ed. R. Ueda (Jpn. Soc.
        Appl. Phys., Tokyo, 1982)   173-176 (1982)

961 Kopp,W., Drummond,T.J., Wang,T., Morkoc,H.,
        Su,S.L.
    A novel camel diode gate GaAs FET
    IEEE Electron Device Lett. EDL-3   86-88 (1982)

962 Kopp,W., Fischer,R., Thorne,R.E., Su,S.L.,
        Drummond,T.J., Morkoc,H., Cho,A.Y.
    A new Al(0.3)Ga(0.7)As/GaAs modulation-doped FET
    IEEE Electron Device Lett. EDL-3   109-111 (1982)

963 Kopp,W., Morkoc,H., Drummond,T.J., Su,S.L.
Characteristics of submicron gate GaAs FET's with
   Al(0.3)Ga(0.7)As buffers: effects of interface
   quality
IEEE Electron Device Lett. EDL-3   46-48 (1982)

964 Kopp,W., Su,S.L., Fischer,R., Lyons,W.G.,
    Thorne,R.E., Drummond,T.J., Morkoc,H., Cho,A.Y.
Use of a GaAs smoothing layer to improve the
   heterointerface of GaAs/Al(x)Ga(1-x)As
   field-effect transistors
Appl. Phys. Lett. 41   563-565 (1982)

965 Kowalczyk,S.P., Schaffer,W.J., Kraut,E.A.,
    Grant,R.W.
Determination of the InAs-GaAs(100) heterojunction
   band discontinuities by x-ray photoelectron
   spectroscopy (XPS)
J. Vac. Sci. Technol. 20   705-708 (1982)

966 Kroemer,H., Zhu,Q.G.
 T  On the interface connection rules for effective-mass
   wave functions at an abrupt heterojunction between
   two semiconductors with different effective mass
J. Vac. Sci. Technol. 21   551-553 (1982)

967 Kubiak,R.A., Driscoll,P., Parker,E.H.C.
A simple source design for MBE
J. Vac. Sci. Technol. 20   252-253 (1982)

968 Kuenzel,H., Doehler,G.H., Ploog,K.
Determination of photoexcited carrier concentration
   and mobility in GaAs doping superlattices by hall
   effect measurements
Appl. Phys. A 27   1-10 (1982)

969 Kuenzel,H., Jung,H., Schubert,E.F., Ploog,K.
Influence of growth conditions and of alloy
   composition on electrical and optical properties
   of MBE Al(x)Ga(1-x)As (0.2 <= x <= 0.4)
J. Physique 43 Colloque C5   C5/175-C5/182 (1982)

970 Kuenzel,H., Knecht,J., Jung,H., Wuenstel,K.,
    Ploog,K.
The effect of arsenic vapour species on electrical
   and optical properties of GaAs grown by molecular
   beam epitaxy
Appl. Phys. A 28   167-173 (1982)

971 Lambert,B., Deveaud,B., Regreny,A., Talalaeff,G.
Impurity photoluminescence in GaAs/Ga(1-x)Al(x)As
   multiple quantum wells
Solid State Commun. 43   443-446 (1982)

972  Landgren,G., Ludeke,R., Serrano,C.M.
     Epitaxial Al films on GaAs(100) surfaces
     J. Cryst. Growth 60   393-402 (1982)

973  Landgren,G., Svensson,S.P., Andersson,T.G.
     Temperature and reconstruction dependence of the
        initial Al growth on GaAs(001)
     Surf. Sci. 122   55-68 (1982)

974  Larsen,P.K., Veen,J.F. van der
     Surface band structure of MBE-grown GaAs(001)-2x4
     J. Phys. C 15   L431-L435 (1982)

975  Larsen,P.K., Veen,J.F. van der, Mazur,A.,
        Pollmann,J., Neave,J.H., Joyce,B.A.
     Surface electronic structure of GaAs(001)-(2x4):
        Angle-resolved photoemission and tight-binding
        calculations
     Phys. Rev. B 26   3222-3237 (1982)

976  Laughlin,R.B.
  T  Impurities and edges in the quantum Hall effect
     Surf. Sci. 113   22-26 (1982)

977  Laviron,M., Delagebeaudeuf,D., Delescluse,P.,
        Etienne,P., Chaplart,J., Linh,N.T.
     Low noise normally on and normally off
        two-dimensional electron gas field effect
        transistors
     Appl. Phys. Lett. 40   530-532 (1982)

978  Levine,B.F., Tsang,W.T., Bethea,C.G., Capasso,F.
     Electron drift velocity measurement in
        compositionally graded Al(x)Ga(1-x)As by
        time-resolved optical picosecond reflectivity
     Appl. Phys. Lett. 41   470-472 (1982)

979  Linh,N.T., Delagebeaudeuf,D., Laviron,M.,
        Delescluse,P., Chaplart,J.
     Low-noise two dimensional electron gas MESFETs
     Inst. Phys. Conf. Ser. 63   585-586 (1982)

980  Linh,N.T., Tung,P.N., Delagebeaudeuf,D.,
        Delescluse,P., Laviron,M.
  R  High speed - low power GaAs/AlGaAs TEGFET integrated
        circuit
     IEDM 82 Technical Digest   582-585 (1982)

981  Liu,Y.Z.
     An enhancement mode Schottky barrier gate
        charge-coupled device on a high electron mobility
        transistor structure
     Appl. Phys. Lett. 41   874-876 (1982)

982 Low,T.S., Stillman,G.E., Cho,A.Y., Morkoc,H., Calawa,A.R.
Spectroscopy of donors in high purity GaAs grown by molecular beam epitaxy
Appl. Phys. Lett. 40    611-613 (1982)

983 Low,T.S., Stillman,G.E., Collins,D.M., Wolfe,C.M., Tiwari,S., Eastman,L.F.
Spectroscopic identification of Si donors in GaAs
Appl. Phys. Lett. 40    1034-1036 (1982)

984 Ludeke,R., Chiang,T.C., Eastman,D.E.
Crystallographic relationships and interfacial properties of Ag on GaAs(100) surfaces
J. Vac. Sci. Technol. 21    599-606 (1982)

985 Lykakh,V.A., Tetervov,A.P.
T  Cyclotron resonance in a degenerate plasma of a semiconductor with a superlattice
Sov. Phys. Semicond. 18    1358-1360    *    Fiz. Tekh. Poluprovodn. 16    2105-2109 (1982)

986 Maan,J.C., Altarelli,M., Sigg,H., Wyder,P., Chang,L.L., Esaki,L.
Effective mass determination of a highly doped InAs-GaSb superlattice using helicon wave propagation
Surf. Sci. 113    347-352 (1982)

987 Maan,J.C., Englert,T., Tsui,D.C., Gossard,A.C.
Observation of cyclotron resonance in the photoconductivity of two-dimensional electrons
Appl. Phys. Lett. 40    609-610 (1982)

988 Maan,J.C., Uihlein,C., Chang,L.L., Esaki,L.
Hybrid cyclotron-intersubband resonance in thin InAs layers confined between GaSb
Solid State Commun. 44    653-656 (1982)

989 Madhukar,A.
T  Modulated semiconductor structures: An overview of some basic considerations for growth and desired electronic structure
J. Vac. Sci. Technol. 20    149-161 (1982)

990 Mailhiot,C., Chang,Y.C., McGill,T.C.
T  Energy spectra of donors in GaAs-Ga(1-x)Al(x)As quantum well structures in the effective-mass approximation
Phys. Rev. B 26    4449-4457 (1982)

991 Mailhiot,C., Chang,Y.C., McGill,T.C.
T  Energy spectra of donors in GaAs-Ga(1-x)Al(x)As quantum well structures in the effective mass approximation
J. Vac. Sci. Technol. 21    519-523 (1982)

992  Mailhiot,C., Chang,Y.C., McGill,T.C.
 T   Energy spectra of donors in GaAs-Ga(1-x)Al(x)As
       quantum well structures
     Surf. Sci. 113   161-164 (1982)

993  Malik,R.J.
     The design and growth of GaAs planar doped barriers
       by MBE
     Mater. Lett. 1   22-25 (1982)

994  Malik,R.J.
 T   Planar doped barriers and their device applications
     Collected Papers of MBE-CST-2, Ed. R. Ueda (Jpn. Soc.
       Appl. Phys., Tokyo, 1982)   29-32 (1982)

995  Malik,R.J.
     Planar doped barriers by molecular beam epitaxy for
       millimeter wave devices
     Proc. SPIE - Int. Soc. Opt. Eng. 317   243-250 (1982)

996  Malik,R.J., Dixon,S.
     A subharmonic mixer using a planar doped barrier
       diode with symmetric conductance
     IEEE Electron Device Lett. EDL-3   205-207 (1982)

997  Maloney,T.J., Saxena,R.R., Chai,Y.G.
     AlGaAs/GaAs JFETs by organometallic and molecular
       beam epitaxy
     Electron. Lett. 18   112-113 (1982)

998  Massies,J., Delescluse,P., Etienne,P., Linh,N.T.
     The growth of silver on GaAs(100): epitaxial
       relationships, mode of growth and interfacial
       diffusion
     Thin Solid Films 90   113-118 (1982)

999  Massies,J., Delescluse,P., Linh,N.T.
 R   The use of MBE and standard surface techniques in the
       study of the growth of metals on GaAs
     Collected Papers of MBE-CST-2, Ed. R. Ueda (Jpn. Soc.
       Appl. Phys., Tokyo, 1982)   287-290 (1982)

1000 Massies,J., Linh,N.T.
     On the growth of silver on GaAs(100) surfaces
     J. Cryst. Growth 56   25-38 (1982)

1001 Massies,J., Linh,N.T.
     Epitaxial relationships between Al, Ag, and GaAs(001)
       surfaces
     Surf. Sci. 114   147-160 (1982)

1002 Massies,J., Linh,N.T.
     Ag M(4,5)N(4,5)N(4,5) Auger lineshape variation
       during the epitaxial growth of Ag onto GaAs(001)
     J. Physique 43   939-944 (1982)

1003  Massies,J., Rochette,J.F., Delescluse,P.,
      Etienne,P., Chevrier,J., Linh,N.T.
      High-mobility Ga(0.47)In(0.53)As thin epitaxial
         layers grown by MBE, very closely lattice-matched
         to InP
      Electron. Lett. 18    758-760 (1982)

1004  Masu,K., Hiroi,S., Mishima,T., Konagai,M.,
      Takahashi,K.
      Preparation of (Al(x)Ga(a-x))(y)In(1-y)As (0<=x<=0.5,
         y=0.47) lattice matched to InP substrate grown by
         molecular beam epitaxy
      Inst. Phys. Conf. Ser. 63    577-578 (1982)

1005  Masu,K., Mishima,T., Hiroi,S., Konagai,M.,
      Takahashi,K.
      Preparation of [Al(x)Ga(1-x)](y)In(1-y)As
         (0<=x<=0.5,y=0.47) lattice matched to InP
         substrates by molecular beam epitaxy
      J. Appl. Phys. 53    7558-7560 (1982)

1006  McAfee,S.R., Lang,D.V., Tsang,W.T.
      Observation of deep levels associated with the
         GaAs/AlxGa1-xAs interface grown by molecular beam
         epitaxy
      Appl. Phys. Lett. 40    520-522 (1982)

1007  McLevige,W.V., Yuan,H.T., Duncan,W.M.,
      Frensley,W.R., Doerbeck,F.H., Morkoc,H.,
      Drummond,T.J.
      GaAs/AlGaAs heterojunction bipolar transistors for
         integrated circuit applications
      IEEE Electron. Device Lett. EDL-3    43-45 (1982)

1008  Mendez,E.E., Bastard,G., Chang,L.L., Esaki,L.,
      Morkoc,H., Fischer,R.
      Effect of an electric field on the luminescence of
         GaAs quantum wells
      Phys. Rev. B 26    7101-7104 (1982)

1009  Mendez,E.E., Chang,L.L., Esaki,L.
  T   Two-dimensional quantum states in
         multi-heterostructures of three constituents
      Surf. Sci. 113    474-478 (1982)

1010  Merlin,R., Pinczuk,A., Beard,W.T., Wood,C.E.C.
      Light scattering study of electrons confined at the
         Ge/GaAs interfaces
      J. Vac. Sci. Technol. 21    516-518 (1982)

1011  Milano,R.A., Cohen,M.J., Miller,D.L.
      Modulation-doped AlGaAs/GaAs heterostructure charge
         coupled devices
      IEEE Electron Devive Lett. EDL-3    194-196 (1982)

1012  Milanovic,V., Tjapkin,D.
  T   Determination of the potential distribution in
         semiconductor heterostructures in the presence of
         quantum effects
      Physica 114B   375-378 (1982)

1013  Milanovic,V., Tjapkin,D.
  T   Energy band calculation and zero energy gap
         conditions for semiconductor superlattices
      Phys. Status Solidi B 110   687-695 (1982)

1014  Miller,D.A.B., Chemla,D.S., Eilenberger,D.J.,
         Smith,P.W., Gossard,A.C., Tsang,W.T.
      Large room-temperature optical nonlinearity in
         GaAs/Ga(1-x)Al(x)As multiple quantum well
         structures
      Appl. Phys. Lett. 41   679-681 (1982)

1015  Miller,D.L., Asbeck,P.M., Petersen,W.C.
      (AlGa)As/GaAs bipolar transistors grown by molecular
         beam epitaxy
      Collected Papers of MBE-CST-2, Ed. R. Ueda (Jpn. Soc.
         Appl. Phys., Tokyo, 1982)   121-124 (1982)

1016  Miller,D.L., Harris,J.S.,Jr., Asbeck,P.M.
      An MBE AlGaAs/GaAs heterojunction bipolar transistor
      Inst. Phys. Conf. Ser. 63   579-580 (1982)

1017  Miller,D.L., Yang,H.T., Zehr,S.W.
      Cascade AlGaAs-GaAs solar cell researching using
         molecular beam epitaxy
      Proc. SPIE - Int. Soc. Opt. Eng. 323   17-22 (1982)

1018  Miller,D.L., Zehr,S.W., Harris,J.S.,Jr.
      GaAs-AlGaAs tunnel junctions for multigap casade
         solar cells
      J. Appl. Phys. 53   2084-2088 (1982)

1019  Miller,R.C., Gossard,A.C., Tsang,W.T., Munteanu,O.
      Extrinsic photoluminescence from GaAs quantum wells
      Phys. Rev. B 25   3871-3877 (1982)

1020  Miller,R.C., Gossard,A.C., Isang,W.T., Munteanu,O.
      Bound excitons in p-doped GaAs quantum wells
      Solid State Commun. 43   519-522 (1982)

1021  Miller,R.C., Kleinman,D.A., Gossard,A.C.,
         Munteanu,O.
      Biexcitons in GaAs quantum wells
      Phys. Rev. B 25   6545-6547 (1982)

1022  Miller,R.C., Tsang,W.T., Munteanu,O.
      Extrinsic layer at Al(x)Ga(1-x)As-GaAs interfaces
      Appl. Phys. Lett. 41   374-376 (1982)

1023 **Mimura,T.**
R The present status of modulation-doped and insulated-gate field-effect transistors in III-V semiconductors
Surf. Sci. 113    454-463 (1982)

1024 **Miyao,M., Chinen,K., Niigaki,M., Hagino,M.**
MBE growth of transmission photocathode and 'in situ' NEA activation
Collected Papers of MBE-CST-2, Ed. R. Ueda (Jpn. Soc. Appl. Phys., Tokyo, 1982)    107-109 (1982)

1025 **Mon,K.K.**
T Electronic band structure of (001)GaAs-AlAs superlattices
Solid State Commun. 41    699-700 (1982)

1026 **Morgan,D.V., Board,K., Wood,C.E.C., Eastman,L.F.**
Current transport in Al/InAlAs/InGaAs heterostructures
Phys. Status Solidi A 72    251-260 (1982)

1027 **Morkoc,H.**
A short-channel GaAs FET fabricated like a MESFET, but operating like a JFET
Jpn. J. Appl. Phys. 21    L233-L234 (1982)

1028 **Morkoc,H.**
Influence of MBE growth conditions on the properties of Al(x)Ga(1-x)As/GaAs heterostructures
J. Physique 43 Colloque C5    C5/209-C5/220 (1982)

1029 **Morkoc,H.**
Short-channel GaAs FET fabricated like a MESFET but operating like a JFET
Electron. Lett. 18    258-259 (1982)

1030 **Morkoc,H., Drummond,T.J., Camras,M.D., Holonyak,N.,Jr.**
Short-wavelength continuous 300-K photopumped $Al_xGa_{1-x}As$-GaAs quantum well heterostructure laser
Appl. Phys. Lett. 40    18-19 (1982)

1031 **Morkoc,H., Drummond,T.J., Fischer,R.**
Interfacial properties of (Al,Ga)As/GaAs structures: Effect of substrate temperature during growth by molecular beam epitaxy
J. Appl. Phys. 53    1030-1033 (1982)

1032 **Morkoc,H., Drummond,T.J., Fischer,R., Cho,A.Y.**
Moderate mobility enhancement in single period $Al_xGa_{1-x}As$/GaAs heterojunctions with GaAs on top
J. Appl. Phys. 53    3321-3323 (1982)

1033 **Morkoc,H., Drummond,T.J., Kopp,W., Fischer,R.**
Influence of substrate temperature on morphology of
   AlxGa1-xAs grown by molecular beam epitaxy
J. Electrochem. Soc. 129   824-826 (1982)

1034 **Morkoc,H., Drummond,T.J., Omori,M.**
GaAs MESFET's by molecular beam epitaxy
IEEE Trans. Electron Devices ED-29   222-224 (1982)

1035 **Morkoc,H., Holonyak,N.,Jr., Drummond,T.J.,
   Camras,M.D., Fischer,R.**
Al(x)Ga(1-x)As/GaAs quantum well heterojunction
   lasers grown by molecular beam epitaxy
Proc. SPIE - Int. Soc. Opt. Eng. 323   13-16 (1982)

1036 **Morkoc,H., Kopp,W., Drummond,T.J., Su,S.L.,
   Thorne,R.E., Fischer,R., Wang,T.**
Submicron gate GaAs/Al0.3Ga0.7AS MESFETs with
   exremely sharp interfaces (40A)
IEEE Trans. Electron Devices ED-29   1013-1018 (1982)

1037 **Morkoc,H., Stamberg,R., Krikorian,E.**
Whisker growth during epitaxy of GaAs by molecular
   beam epitaxy
Jpn. J. Appl. Phys. 21   L230-L232 (1982)

1038 **Munoz-Yague,A., Baceiredo,S.**
Ge incorporation in GaAs grown by molecular beam
   epitaxy: A thermodynamic study
J. Electrochem. Soc. 129   2108-2113 (1982)

1039 **Muro,K., Narita,S., Hiyamizu,S., Nanbu,K.,
   Hashimoto,H.**
Far-infrared cyclotron resonance of two-dimensional
   electrons in an Al(x)Ga(1-x)As/GaAs heterojunction
Surf. Sci. 113   321-325 (1982)

1040 **Murschall,R., Gant,H., Moench,W.**
Low-energy electron energy-loss spectroscopy with
   Ge:GaAs(110) heterostructures
Solid State Commun. 42   787-791 (1982)

1041 **Naganuma,M., Suzuki,Y., Okamoto,H.**
Photoluminescence of GaSb-AlSb superlattices grown by
   MBE
Inst. Phys. Conf. Ser. 63   125-130 (1982)

1042 **Narita,S., Takeyama,S., Luo,W., Hiyamizu,S.,
   Nanbu,K., Hashimoto,H.**
Quantum galvanomagnetic properties of two-dimensional
   electron gas in AlxGa1-xAs/GaAs heterojunction FET
   in strong magnetic fields
Surf. Sci. 113   301-305 (1982)

1043 **Nathan,M.I., Heiblum,M.**
An improved AuGe ohmic contact to n-GaAs
Solid-State Electron. 25   1063-1065 (1982)

1044 **Noreika,A.J., Takei,W.J., Francombe,M.H.,
Wood,C.E.C.**
Indium antimonide-bismuth compositions grown by
molecular beam epitaxy
J. Appl. Phys. 53   4932-4937 (1982)

1045 **Noreika,A.J., Takei,W.J., Francombe,M.H.,
Wood,C.E.C.**
Structure and electrical properties of MBE-grown
In-Sb based compositions
Collected Papers of MBE-CST-2, Ed. R. Ueda (Jpn. Soc.
Appl. Phys., Tokyo, 1982)   161-164 (1982)

1046 **Norrman,S.H., Andersson,T.G., Svensson,S.P.,
Flemming,K.E.**
Highly stabilised evaporation sources in a
water-cooled carousel housing
J. Phys. E 15   731-735 (1982)

1047 **O'Clock,G.D.,Jr., Erickson,L.P.**
R   Impact of molecular beam epitaxy on millimeter wave
and optical systems
Proc. SPIE - Int. Soc. Opt. Eng. 317   268-274 (1982)

1048 **O'Connell,R.F.**
R   Two dimensional systems in solid state and surface
physics: strong electric and magnetic fields
effects
J. Physique 43, Colloque C2   C2/81-C2/96 (1982)

1049 **O'Connor,P., Pearsall,T.P., Cheng,K.Y., Cho,A.Y.,
Hwang,J.C.M., Alavi,K.**
In0.53Ga0.47As FET's with insulator-assisted Schottky
gates
IEEE Electron. Device Lett. EDL-3   64-66 (1982)

1050 **Ogura,M., Yao,T., Hata,T.**
T   Surface emitting laser diode with multilayered
heterostructure reflectors
Collected Papers of MBE-CST-2, Ed. R. Ueda (Jpn. Soc.
Appl. Phys., Tokyo, 1982)   69-72 (1982)

1051 **Ohno,H., Barnard,J.A.**
R   Field-effect transistors
in "GaInAsP Alloy Semiconductors", Ed. T.P. Pearsall
(Wiley, Chichester, UK, 1982)   437-455 (1982)

1052 **Ohno,H., Sakaki,H.**
Tangential magnetoresistance of two-dimensional
electron gas at a selectively doped n-GaAlAs/GaAs
heterojunction interface grown by molecular beam
epitaxy
Appl. Phys. Lett. 40   893-895 (1982)

1053 **Ohno,H., Sakaki,H.**
Tangential magnetoresistivity of two dimensional
electron gas at a selectively doped n-GaAsAs/GaAs
heterojunction
Collected Papers of MBE-CST-2, Ed. R. Ueda (Jpn. Soc.
Appl. Phys., Tokyo, 1982)   135-138 (1982)

1054 **Okamoto,K., Wood,C.E.C., Rathbun,L., Eastman,L.F.**
Oxygen stabilization of molecular beam epitaxial
Al-GaAs Schottky barrier heights
J. Appl. Phys. 53   4521-4523 (1982)

1055 **Okamoto,K., Wood,C.E.C., Rathbun,L., Eastman,L.F.**
Instabilities in the growth of
AlxGa(1-x)As/Al/AlyGa(1-y)As structures by
molecular beam epitaxy
J. Appl. Phys. 53   1532-1535 (1982)

1056 **Olego,D., Chang,T.Y., Silberg,E., Caridi,E.A., Pinczuk,A.**
Compositional dependence of band-gap energy and
conduction-band effective mass of
In(1-x-y)Ga(x)Al(y)As lattice matched to InP
Appl. Phys. Lett. 41   476-478 (1982)

1057 **Olego,D., Pinczuk,A., Gossard,A.C., Wiegmann,W.**
Plasma dispersion in a layered electron gas: A
determination in GaAs-(AlGa)As heterostructures
Phys. Rev. B 25   7867-7870 (1982)

1058 **Olego,D., Pinczuk,A., Gossard,A.C., Wiegmann,W.**
Plasma dispersion in a layered electron gas: a
determination in GaAs-(AlGa)As heterostructures
Phys. Rev. B 25   7867-7870 (1982)

1059 **Ono,Y.**
T   Self-consistent treatment of dynamical diffusion
coefficient of two dimensional random electron
system under strong magnetic fields
J. Phys. Soc. Jpn. 51   3544-3552 (1982)

1060 **Ono,Y.**
T   Energy dependence of localization length of
two-dimensional electron system moving in a random
potential under strong magnetic fields
J. Phys. Soc. Jpn. 51   2055-2056 (1982)

**1061 Ono,Y.**
T  A calculation of two-dimensional Hall conductivity under strong magnetic fields — Wigner representation
in "Anderson Localization", Eds. Y. Nagaoka and H. Fukuyama, Springer Ser. Solid-State Sci. Vol. 39 207-215 (1982)

**1062 Paalanen,M.A., Tsui,D.C., Gossard,A.C.**
Quantized Hall effect at low temperatures
Phys. Rev. B25   5566-5569 (1982)

**1063 Palmier,J.F., Chomette,A.**
T  Phonon-limited near equilibrium transport in a semiconductor superlattice
J. Physique 43   381-391 (1982)

**1064 Parker,E.H.C., King,R.M.**
R  New channels for microchips
New Sci. 96 No. 1327   105-108 (1982)

**1065 Petroff,P.M., Cho,A.Y., Reinhart,F.K., Gossard,A.C., Wiegmann,W.**
Alloy clustering in $Ga_{1-x}Al_xAs$ compound semiconductors grown by molecular beam epitaxy
Phys. Rev. Lett. 48   170-173 (1982)

**1066 Petroff,P.M., Gossard,A.C., Logan,R.A., Wiegmann,W.**
Toward quantum well wires: Fabrication and optical properties
Appl. Phys. Lett. 41   635-638 (1982)

**1067 Pinczuk,A., Worlock,J.M.**
R  Light scattering by two-dimensional electron systems in semiconductors
Surf. Sci. 113   69-84 (1982)

**1068 Ploog,K.**
R  Molecular beam epitaxy of III-V compounds: application of MBE-grown films
Ann. Rev. Mater. Sci. 12   123-148 (1982)

**1069 Ploog,K.**
R  GaAs doping superlattices — a new class of semiconductor materials grown by molecular beam epitaxy
Collected Papers of MBE-CST-2, Ed. R. Ueda (Jpn. Soc. Appl. Phys., Tokyo, 1982)   17-20 (1982)

**1070 Ploog,K., Kuenzel,H.**
R  Growth and properties of new artificial doping superlattices in GaAs
Microelectron. J. 13 No. 3   5-22 (1982)

1071    Ploog,K., Kuenzel,H., Collins,D.M.
        Comment on "A photoluminescence study of
            beryllium-doped GaAs grown by molecular beam
            epitaxy"
        J. Appl. Phys. 53    6467-6468 (1982)

1072    Portal,J.C., Nicholas,R.J., Brummell,M.A.,
            Cho,A.Y., Cheng,K.Y., Pearsall,T.P.
        Quantum transport in GaInAs-AlInAs heterojunctions,
            and the influence of intersubband scattering
        Solid State Commun. 43    907-911 (1982)

1073    Prange,R.E., Joynt,R.
  T     Conduction in a strong field in two dimensions: the
            quantum Hall effect
        Phys. Rev. B 25    2945-2946 (1982)

1074    Price,G.L.
        Preservation and regeneration of an MBE grown surface
        Collected Papers of MBE-CST-2, Ed. R. Ueda (Jpn. Soc.
            Appl. Phys., Tokyo, 1982)    259-262 (1982)

1075    Price,P.J.
  TR    Electron transport in polar heterolayers
        Surf. Sci. 113    199-210 (1982)

1076    Price,P.J.
  T     Hot electrons in GaAs heterolayer at low temperature
        J. Appl. Phys. 53    6863-6866 (1982)

1077    Quinn,J.J., McCombe,B.D.
  R     The IVth international conference on the electronic
            properties of two-dimensional systems (EP2DS-IV)
        Comments Solid State Phys. 10    139-154 (1982)

1078    Reich,R.K., Ferry,D.K.
  T     Moment equations in the Wigner formulation for
            superlattice band structures
        Phys. Lett. A 91    31-32 (1982)

1079    Reich,R.K., Grondin,R.O., Ferry,D.K., Iafrate,G.J.
  T     Transport in surface superlattices
        Phys. Lett. A 91    28-30 (1982)

1080    Reich,R.K., Grondin,R.O., Ferry,D.K., Iafrate,G.J.
  T     The Bloch-FET - a lateral surface superlattice device
        IEEE Electron Device Lett. EDL-3    381-383 (1982)

1081    Ridley,B.K.
  T     The electron-phonon interaction in
            quasi-two-dimensional semiconductor quantum-well
            structures
        J. Phys. C 15    5899-5917 (1982)

1082 **Rocket,A., Drummond,T.J., Greene,J.E., Morkoc,H.**
Surface segregation model for Sn-doped GaAs grown by molecular beam epitaxy
J. Appl. Phys. 53    7085-7087 (1982)

1083 **Sakaki,H.**
T   Physical limits and applications of extremely high electron mobility effects in ultrafine semiconductor wire (USW) structures
Inst. Phys. Conf. Ser. 63    251-256 (1982)

1084 **Sakaki,H.**
T   Velocity-modulation transistor (VMT) - a new field-effect transistor concept
Jpn. J. Appl. Phys. 21    L381-L383 (1982)

1085 **Sakamoto,T.**
R   Moving toward future electron devices
JEE, J. Electron. Eng. 19 No.7    28-31 (1982)

1086 **Sanchez-Dehesa,J., Tejedor,C.**
T   Self-consistent calculation of properties of GaAs-AlAs superlattices with homopolar interfaces
Phys. Rev. B 26    5824-5831 (1982)

1087 **Sauvage,M., Massies,J.**
Characterization of MBE grown silver layers on (001)GaAs surfaces by synchrotron radiation plane wave techniques
J. Cryst. Growth 59    605-615 (1982)

1088 **Scott,G.B., Dobson,P.J., Duggan,G., Dawson,P.**
Reply to "Comment on 'A photoluminescence study of beryllium-doped GaAs grown by molecular beam epitaxy'"
J. Appl. Phys. 53    6469-6470 (1982)

1089 **Sekiguchi,Y., Sakaki,H., Tanoue,T., Hotta,T., Ohno,H.**
Transport properties of electrons at n-AlGaAs/GaAs heterojunction interface and their dependence on GaAs buffer-layer thickness and substrates
Collected Papers of MBE CST 1, Ed. K. Ueda (Jpn. Soc. Appl. Phys., Tokyo, 1982)    139-142 (1982)

1090 **Shank,C.V., Fork,R.L., Greene,B.I., Weisbuch,C., Gossard,A.C.**
Picosecond dynamics of highly excited multiquantum well structures
Surf. Sci. 113    108-111 (1982)

1091 **Shmelev,G.M., Enaki,N.A.**
T   Conductivity of a superlattice during electron scattering by optical phonons
Sov. Phys. Journal 25    78-91    *    Izv. Vyssh. Zaved., Fiz., 25    81-84 (1982)

1092　Singh,J., Madhukar,A.
 T　Monte Carlo simulation of the growth of A(1-x)B(x) layers on lattice matched substrates in molecular beam epitaxy
　　J. Vac. Sci. Technol. 20　716-719 (1982)

1093　Smith,R.S., Ganser,P.M., Ennen,H.
　　Selenium doping of molecular beam epitaxial GaAs using SnSe2
　　J. Appl. Phys. 53　9210-9211 (1982)

1094　Spicer,W.E., Eglash,S.J., Skeath,P., Mahowald,P., Lindau,I., Pan,S., Mo,D., Collins,D.M.
 R　Surface and interface electronic structure and Schottky barriers
　　Collected Papers of MBE-CST-2, Ed. R. Ueda (Jpn. Soc. Appl. Phys., Tokyo, 1982)　269-278 (1982)

1095　Stagg,J.P., Hulyer,P.J., Foxon,C.T., Ashenford,D.
　　Optimisation of Sn doping in GaAlAs/GaAs DH lasers grown by MBE
　　J. Physique 43 Colloque C5　C5/377-C5/384 (1982)

1096　Stoermer,H.L., Gossard,A.C., Wiegmann,W.
　　Observation of intersubband scattering in a 2-dimensional electron system
　　Solid State Commun. 41　707-709 (1982)

1097　Stoermer,H.L., Tsui,D.C., Gossard,A.C.
 R　Zero resistance state and origin of the quantized Hall effect in two-dimensional electron systems
　　Surf. Sci. 113　32-38 (1982)

1098　Streda,P.
 T　Theory of quantised Hall conductivity in two dimensions
　　J. Phys. C 15　L717-L721 (1982)

1099　Streda,P.
 T　Quantised Hall effect in a two-dimensional periodic potential
　　J. Phys. C 15　L1299-L1303 (1982)

1100　Su,S.L., Fischer,R., Drummond,T.J., Lyons,W.G., Thorne,R.E., Kopp,W., Morkoc,H.
　　Modulation-doped (Al,Ga)As/GaAs FETs with high transconductance and electron velocity
　　Electron. Lett. 18　794-796 (1982)

1101　Su,S.L., Thorne,R.E., Fischer,R., Lyons,W.G., Morkoc,H.
　　Influence of buffer thickness on the performance of GaAs field effect transistors prepared by molecular beam epitaxy
　　J. Vac. Sci. Technol. 21　961-964 (1982)

1102 Sugahara,T., Wada,O., Fujii,T., Hiyamizu,S., Sakurai,T.
Monolithic 1x4 array of uniform radiance AlGaAs-GaAs LED's grown by molecular beam epitaxy
Jpn. J. Appl. Phys. 21  L349-L350 (1982)

1103 Sugai,S., Harris,J.H., Nurmikko,A.V.
Strong electron-phonon interaction effects in modulated transient reflectance spectra of Ga(0.50)In(0.50)P
Solid State Commun. 43  913-916 (1982)

1104 Sugiura,H., Yamaguchi,M., Yamamoto,A., Shibukawa,A., Uemura,C.
ESCA studies of InP substrate surface thermally cleaned under arsenic molecular beam exposure for MBE growth
Collected Papers of MBE-CST-2, Ed. R. Ueda (Jpn. Soc. Appl. Phys., Tokyo, 1982)  255-258 (1982)

1105 Sugiyama,K.
Molecular beam epitaxy of InSb films on CdTe
J. Cryst. Growth 60  450-452 (1982)

1106 Sullivan,P.W., Farrow,R.F.C., Jones,G.R.
Insulating epitaxial films of BaF2, CaF2, and Ba(x)Ca(1-x)F2 grown by MBE on InP substrates
J. Cryst. Growth 60  403-413 (1982)

1107 Sun,D.C., Sakaki,H., Ohno,H., Sekiguchi,Y., Tanoue,T.
Stabilization of Schottky barrier properties of single-crystal Al/GaAs and Al/AlGaAs/GaAs contacts prepared by molecular beam epitaxy
Inst. Phys. Conf. Ser. 63  311-316 (1982)

1108 Suzuki,Y., Okamoto,H.
Refractive index of GaAs-AlAs superlattice grown by MBE
Collected Papers of MBE-CST-2, Ed. R. Ueda (Jpn. Soc. Appl. Phys., Tokyo, 1982)  51-54 (1982)

1109 Swaminathan,V., Zilko,J.L., Tsang,W.T., Wagner,W.R.
Photoluminescence study of acceptors in Al(x)Ga(1-x)As
J. Appl. Phys. 53  5163-5168 (1982)

1110 Tabatabaie-Alavi,K., Choudhury,A.N.M.M., Alavi,K., Vlcek,J., Slater,N.J., Fonstad,C.G., Cho,A.Y.
Ion-implanted In(0.53)Ga(0.47)As/In(0.52)Al(0.48)As lateral PNP transistors
IEEE Electron Device Lett. EDL-3  379-381 (1982)

1111 Tabatabaie-Alavi,K., Choudhury,A.N.M.M., Alavi,K., Vlcek,J., Slater,N.J., Fonstad,C.G., Cho,A.Y.
(In,Ga)As/(In,Al,)As heterojunction lateral pnp transistors
IEDM 82 Technical Digest    766-769 (1982)

1112 Tacano,M., Sugiyama,Y., Ogura,M., Kawashima,M.
Fabrication of super-Schottky diode by MBE growth and subsequent metal deposition
Collected Papers of MBE-CST-2, Ed. R. Ueda (Jpn. Soc. Appl. Phys., Tokyo, 1982)    125-128 (1982)

1113 Takeda,Y., Kamei,H., Sasaki,A.
T Mobility calculation of two-dimensional electron gas in GaAs/AlGaAs heterostructure at 4.2 K
Electron. Lett. 18    309-311 (1982)

1114 Tanoue,T., Sakaki,H.
Long-lifetime photoconductivity in selectively doped n-AlGaAs/GaAs heterostructures
Collected Papers of MBE-CST-2, Ed. R. Ueda (Jpn. Soc. Appl. Phys., Tokyo, 1982)    143-146 (1982)

1115 Tanoue,T., Sakaki,H.
T A new method to control impact ionization rate ratio by spatial separation of avalanching carriers in multilayered heterostructures
Appl. Phys. 41    67-69 (1982)

1116 Tarng,S.S., Tai,K., Jewell,J.L., Gibbs,H.M., Gossard,A.C., McCall,S.L., Passner,A., Venkatesan,T.N.C., Wiegmann,W.
External off and on switching of a bistable optical device
Appl. Phys. Lett. 40    205-207 (1982)

1117 Tarucha,S., Ishibashi,T., Okamoto,H.
Spontaneous and stimulated photoluminescence properties of GaAs-Al(x)Ga(1-x)As multi-quantum-well heterostructures
Collected Papers of MBE-CST-2, Ed. R. Ueda (Jpn. Soc. Appl. Phys., Tokyo, 1982)    43-46 (1982)

1118 Thorne,R.E., Drummond,T.J., Lyons,W.G., Fischer,R., Morkoc,H.
An explanation for anomalous donor activation energies in Al(0.35)Ga(0.55)As
Appl. Phys. Lett. 41    189-191 (1982)

1119 Thorne,R.E., Fischer,R., Su,S.L., Kopp,W., Drummond,T.J., Morkoc,H.
Performance of inverted structure modulation doped Schottky barrier field effect transistor
Jpn. J. Appl. Phys. 21    L223-L224 (1982)

1120 Thorne,R.E., Su,S.L., Kopp,W., Fischer,R., Drummond,T.J., Morkoc,H.
Normally-on and normally-off camel diode gate GaAs field effect transistors for large scale integration
J. Appl. Phys. 53   5951-5958 (1982)

1121 Thouless,D.J.
T  Localization in a strong magnetic field
in "Anderson Localization", Eds. Y. Nagaoka and H. Fukuyama, Springer Ser. Solid-State Sci. Vol. 39 191-197 (1982)

1122 Thouless,D.J., Kohmoto,M., Nightingale,M.P., Nijs,M. den
T  Quantized Hall conductance in a two-dimensional periodic potential
Phys. Rev. Lett. 49   405-408 (1982)

1123 Tien,Z.J., Perry,C.H., Worlock,J.M., Pinczuk,A., Aggarwal,R.
Resonant Raman scattering from electrons in gallium arsenide-aluminium gallium arsenide heterostructures in high magnetic fields
Proc. 8th Int. Conf. Raman Spectrosc., Eds. J. Lascombe and P.V. Huong (Wiley, Chichester, UK, 1982)   441-442 (1982)

1124 Tien,Z.J., Worlock,J.M., Perry,C.H., Pinczuk,A., Aggarwal,R.L., Stoermer,H.L., Gossard,A.C., Wiegmann,W.
Light scattering from two-dimensional electron systems in strong magnetic fields
Surf. Sci. 113   89-93 (1982)

1125 Tsang,W.T.
R  Recent progress in growing reliable (AlGa)As DH lasers by molecular beam epitaxy for optical communication systems
J. Cryst. Growth 56   464-474 (1982)

1126 Tsang,W.T.
Extremely low threshold (AlGa)As graded-index waveguide separate-confinement heterostructure lasers grown by molecular beam epitaxy
Appl. Phys. Lett. 40   217-219 (1982)

1127 Tsang,W.T.
R  Recent developments in MBE lasers and photodetectors
Collected Papers of MBE-CST-2, Ed. R. Ueda (Jpn. Soc. Appl. Phys., Tokyo, 1982)   75-79 (1982)

1128 Tsang,W.T.
R  Novel optoelectronic devices prepared by molecular beam epitaxy
Proc. SPIE - Int. Soc. Opt. Eng. 317   66-73 (1982)

1129 Tsang,W.T., Logan,R.A.
(AlGa)As strip buried-heterostructure lasers prepared by hybrid crystal growth
Electron. Lett. 18   397-398 (1982)

1130 Tsang,W.T., Logan,R.A., Ditzenberger,J.A.
Stripe-geometry laser with in-situ ohmic contact and self-aligned native surface oxide mask for current isolation prepared by molecular beam epitaxy
Electron. Lett. 18   123-124 (1982)

1131 Tsang,W.T., Miller,R.C., Capasso,F., Bonner,W.A.
High quality InP grown by molecular beam epitaxy
Appl. Phys. Lett. 41   467-469 (1982)

1132 Tsang,W.T., Reinhart,F.K., Ditzenberger,J.A.
1.3 um wavelength GaInAsP/InP double heterostructure lasers grown by molecular beam epitaxy
Appl. Phys. Lett. 41   1094-1096 (1982)

1133 Tsang,W.T., Reinhart,F.K., Ditzenberger,J.A.
Molecular-beam epitaxially grown 1.3um GaInAsP/InP double-heterostructure lasers
Electron. Lett. 18   785-786 (1982)

1134 Tsui,D.C., Gossard,A.C., Field,B.F., Cage,M.E., Dziuba,R.F.
Determination of the fine-structure constant using GaAs-Al$_x$Ga$_{1-x}$As heterostructures
Phys. Rev. Lett. 48   3-6 (1982)

1135 Tsui,D.C., Stoermer,H.L., Gossard,A.C.
Two-dimensional magnetotransport in the extreme quantum limit
Phys. Rev. Lett. 48   1559-1561 (1982)

1136 Tsui,D.C., Stoermer,H.L., Gossard,A.C.
Zero-resistance state of two-dimensional electrons in a quantizing magnetic field
Phys. Rev. B25   1405-1407 (1982)

1137 Tung,P.N., Delagebeaudeuf,D., Laviron,M., Delescluse,P., Chaplart,J., Linh,N.T.
High-speed two-dimensional electron-gas FET logic
Electron. Lett. 18   109-110 (1982)

1138 Tung,P.N., Delescluse,P., Delagebeaudeuf,D., Laviron,M., Chaplart,J., Linh,N.T.
High-speed low-power DCFL using planar two-dimensional electron gas FET technology
Electron. Lett. 18   517-519 (1982)

1139 Vechten,J.A. van
T    Intermixing of an AlAs-GaAs superlattice by Zn diffusion
J. Appl. Phys. 53   7082-7084 (1982)

1140 Veen,J.F. van der, Smit,L., Larsen,P.K., Neave,J.H., Joyce,B.A.
Interface electronic structure of Pb on GaAs(001)
J. Vac. Sci. Technol. 21    375-379 (1982)

1141 Vitlina,R.Z., Chaplik,A.V.
T Nonequilibrium plasmons in a two-dimensional electron gas
Sov. Phys. JETP 56    839-842    *    Zh. Eksp. Teor. Fiz. 83    1457-1463 (1982)

1142 Vodjdani,N., Lemarchand,A., Paradan,H.
Parametric studies of GaAs growth by metalorganic molecular beam epitaxy
J. Physique 43 Colloque C5    C5/339-C5/349 (1982)

1143 Voos,M.
R Far-infrared magneto-absorption in InAs-GaSb superlattices
Surf. Sci. 113    94-101 (1982)

1144 Wada,O., Sanada,T., Sakurai,T.
Monolithic integration of an AlGaAs/GaAs DH LED with a GaAs FET driver
IEEE Electron Device Lett. EDL-3    305-307 (1982)

1145 Wagner,W.R., Cho,A.Y.
$Al_{0.3}Ga_{0.7}P_{0.01}As_{0.99}$/GaAs laser heterostructures grown by molecular beam epitaxy
J. Appl. Phys. 53    6032-6036 (1982)

1146 Wang,S.Y., Bloom,D.M., Collins,D.M.
GaAs Schottky photodiode with 3 dB bandwidth of 20 GHz
IEDM 82 Technical Digest    521-524 (1982)

1147 Wang,W.I.
Mobility enhancement in modulation-doped GaAs/AlAs heterostructures grown by molecular beam epitaxy
Appl. Phys. Lett. 41    540-542 (1982)

1148 White,S.R., Marques,G.E., Sham,L.J.
T Effective-mass theory for electrons in heterostructures
J. Vac. Sci. Technol. 21    544-547 (1982)

1149 Wicks,G.W.
R Thin film growth by molecular beam epitaxy
Proc. SPIE - Int. Soc. Opt. Eng. 346    19-24 (1982)

1150 Williams,G.F., Capasso,F., Tsang,W.T.
T The graded bandgap multilayer avalanche photodiode: a new low-noise detector
IEEE Electron Device Lett. ED2-3    71-73 (1982)

1151  Williams,R.S., Paine,B.M., Schaffer,W.J., Kowalczyk,S.P.
      Channeling measurements of lattice disorder at the GaAs-InAs(100) heterojunction
      J. Vac. Sci. Technol. 21    386-388  (1982)

1152  Wood,C.E.C.
   R  MBE doping processes. A review of current understanding
      Collected Papers of MBE-CST-2, Ed. R. Ueda (Jpn. Soc. Appl. Phys., Tokyo, 1982)   153-159  (1982)

1153  Wood,C.E.C.
   R  Molecular beam epitaxy for microwave field effect transitors
      in "GaAs FET Principles and Technology" Eds. J.V. DiLorenzo and D.D.Khandelwal (Artech House, Dedham, Mass., 1982)   101-114  (1982)

1154  Wood,C.E.C.
   R  III-V alloy growth by molecular beam epitaxy
      in "GaInAs Alloy Semiconductors" Ed. T.P.Pearsall (Wiley, Chichester, UK, 1982)   87-106  (1982)

1155  Wood,C.E.C., DeSimone,D., Singer,K., Wicks,G.W.
      Magnesium- and calcium-doping behavior in molecular-beam epitaxial III-V compounds
      J. Appl. Phys. 53    4230-4235  (1982)

1156  Wood,C.E.C., Eastman,L.F., Board,K., Singer,K., Malik,R.J.
      Regenerative switching device using MBE-grown gallium arsenide
      Electron. Lett. 18    676-677  (1982)

1157  Wood,C.E.C., Morgan,D.V., Rathbun,L.
      Molecular-beam epitaxial group III arsenide alloys: Effect of substrate temperature on composition
      J. Appl. Phys. 53    4524-4526    (1982)

1158  Woodbridge,K., Gowers,J.P., Joyce,B.A.
      Structural properties and composition control of GaAs(y)P(1-y) grown by MBE on VPE GaAs(0.63)P(0.37) substrates
      J. Cryst. Growth 60    21-28  (1982)

1159  Worlock,J.M.
   R  Electrons in novel two-dimensional structures
      Nature 297    360-361  (1982)

1160  Wright,S.L., Inada,M., Kroemer,H.
      Polar-on-nonpolar epitaxy: Sublattice ordering in the nucleation and growth of GaP on Si(211) surfaces
      J. Vac. Sci. Technol. 21    534-539  (1982)

1161 Wright,S.L., Kroemer,H.
Operational aspects of a gallium phosphide source of
P2 vapor in molecular beam epitaxy
J. Vac. Sci. Technol. 20   143-148 (1982)

1162 Xin,S.H., Schaff,W.J., Wood,C.E.C., Eastman,L.F.
Capped versus capless heat treatment of molecular
beam epitaxial GaAs
Appl. Phys. Lett. 41   742-744 (1982)

1163 Xin,S.H., Wood,C.E.C., DeSimone,D., Palmateer,S.C.,
Eastman,L.F.
1.40 eV emission band in GaAs
Electron. Lett. 18   3-5 (1982)

1164 Yamakoshi,S., Wada,O., Fujii,T., Hiyamizu,S.,
Sakurai,T.
Ridge-waveguide GaAs/AlGaAs DH lasers grown by MBE
Collected Papers of MBE-CST-2, Ed. R. Ueda (Jpn. Soc.
Appl. Phys., Tokyo, 1982)   89-92 (1982)

1165 Yamakoshi,S., Wada,O., Fujii,T., Hiyamizu,S.,
Sakurai,T.
High performance ridge-waveguide AlGaAs/GaAs
multiquantum-well lasers grown by molecular beam
epitaxy
IEDM 82 Technical Digest   342-345 (1982)

1166 Yata,M., Niwa,K., Ueda,R.
Growth kinetics of InSb thin films by Sb4 and In1
molecular beams
Collected Papers of MBE-CST-2, Ed. R. Ueda (Jpn. Soc.
Appl. Phys., Tokyo, 1982)   249-252 (1982)

1167 Yoshida,S., Misawa,S., Gonda,S.
Reactive molecular beam epitaxy of $Al(x)Ga(1-x)N$
Collected Papers of MBE-CST-2, Ed. R. Ueda (Jpn. Soc.
Appl. Phys., Tokyo, 1982)   165-168 (1982)

1168 Yoshida,S., Misawa,S., Gonda,S.
Properties of $Al(x)Ga(1-x)N$ films prepared by
reactive molecular beam epitaxy
J. Appl. Phys. 53   6844-6848 (1982)

1169 Zeller,C., Abstreiter,G., Ploog,K.
The influence of temperature and incident light
intensity on single particle and collective
excitations in multilayer stuctures
Surf. Sci. 113   85-88 (1982)

1170 Zeller,C., Vinter,B., Abstreiter,G., Ploog,K.
Quasi-two-dimensional photoexcited carriers in GaAs
doping superlattices
Phys. Rev. B 26   2124-2132 (1982)

1171    Zhou,B.L., Ploog,K., Gmelin,E., Zheng,X.Q.,
           Schulz,M.
        Assessment of persistent-photoconductivity centers in
           MBE grown Al(x)Ga(1-x)As using capacitance
           spectroscopy measurements
        Appl. Phys. A 28    223-227 (1982)

1172    Ziel,J.P. van der, Tsang,W.T.
        Integrated multilayer GaAs lasers separated by tunnel
           junctions
        Appl. Phys. Lett. 41    499-501 (1982)

# Subject Categories and References
# Year 1983

**AlAs, GaAs, (AlGa)As**
1173,1174,1175,1179,1183,1184,1185,1186,1187,1188,1189,
1190,1191,1192,1194,1195,1196,1197,1198,1199,1200,1201,
1202,1203,1204,1205,1208,1209,1210,1211,1212,1213,1214,
1215,1216,1217,1218,1219,1220,1221,1222,1223,1224,1225,
1226,1227,1228,1229,1230,1231,1233,1234,1235,1236,1237,
1238,1239,1241,1242,1243,1244,1245,1247,1248,1249,1250,
1251,1252,1253,1254,1255,1256,1257,1258,1259,1261,1262,
1264,1265,1266,1267,1268,1274,1275,1276,1277,1278,1279,
1280,1281,1282,1286,1288,1289,1290,1291,1292,1293,1294,
1295,1296,1297,1298,1299,1300,1301,1302,1303,1304,1305,
1306,1307,1308,1309,1310,1312,1313,1314,1315,1316,1317,
1318,1319,1320,1321,1322,1323,1324,1325,1327,1329,1330,
1331,1332,1333,1334,1335,1337,1340,1341,1342,1343,1344,
1345,1347,1348,1349,1350,1351,1352,1353,1354,1355,1356,
1357,1358,1360,1361,1363,1365,1366,1367,1368,1369,1372,
1373,1374,1375,1376,1377,1378,1379,1380,1381,1382,1383,
1384,1386,1388,1392,1393,1394,1395,1396,1398,1399,1400,
1401,1402,1403,1404,1405,1406,1407,1408,1409,1410,1411,
1412,1413,1414,1415,1416,1417,1418,1419,1420,1421,1422,
1423,1424,1425,1426,1427,1428,1429,1430,1431,1432,1433,
1434,1435,1436,1437,1438,1439,1440,1441,1442,1443,1444,
1445,1446,1447,1448,1449,1450,1451,1453,1454,1455,1456,
1457,1458,1459,1460,1461,1463,1464,1467,1469,1470,1471,
1472,1473,1474,1475,1476,1477,1478,1479,1480,1481,1482,
1483,1484,1485,1486,1487,1488,1489,1490,1491,1492,1493,
1494,1497,1498,1500,1501,1502,1503,1505,1506,1508,1509,
1510,1511,1512,1513,1517,1521,1522,1523,1524,1525,1526,
1527,1528,1530,1531,1532,1533,1534,1535,1536,1537,1538,
1539,1541,1542,1543,1544,1545,1546,1547,1548,1549,1550,
1551,1552,1553,1554,1555,1556,1557,1558,1559,1560,1561,
1562,1564,1568,1569,1571,1572,1573,1574,1575,1576,1578,
1579,1580,1581,1582,1583,1584,1585,1586,1587,1588,1589,
1590,1592,1593,1594,1595,1596,1597,1599,1600,1601,1602,
1604,1605,1606,1607,1609,1610,1613,1614,1615,1616,1617,
1620,1621,1622,1623,1624,1626,1627,1628,1629,1631,1632,
1633,1634,1635,1636,1637,1638,1639,1640,1643,1644,1646,
1647,1648,1649,1650,1651,1652,1653,1654,1655,1656,1657,
1660,1661,1662,1663,1665,1666,1667,1668,1669,1670,1671,
1672,1673

## GaSb, InAs, Ga(AsSb), (GaIn)As
1176,1177,1180,1181,1182,1193,1206,1207,1217,1224,1226,
1232,1237,1238,1239,1240,1244,1246,1249,1252,1260,1263,
1268,1272,1280,1283,1297,1321,1322,1323,1324,1326,1330,
1336,1338,1340,1345,1346,1370,1371,1385,1387,1390,1397,
1405,1449,1452,1462,1464,1465,1468,1495,1496,1507,1508,
1515,1516,1521,1523,1525,1529,1540,1554,1555,1565,1566,
1570,1598,1603,1611,1613,1630,1641,1642,1646,1647,1649,
1657,1671

## InP, (AlIn)As, (GaIn)As, (GaIn)(PAs)
1176,1177,1178,1193,1207,1217,1226,1232,1237,1238,1240,
1246,1260,1263,1267,1268,1269,1270,1271,1283,1297,1321,
1322,1323,1324,1328,1338,1364,1370,1371,1385,1387,1389,
1390,1406,1462,1464,1465,1495,1496,1504,1507,1508,1514,
1515,1516,1518,1519,1520,1521,1523,1540,1555,1565,1566,
1570,1577,1603,1608,1612,1618,1619,1641,1642,1646,1647,
1649,1657,1671

## Growth Equipment
1184,1236,1268,1270,1309,1337,1369,1372,1407,1437,1466,
1523,1567,1619,1647,1659

## Growth Mechanism, Surface Analysis
1185,1187,1196,1197,1198,1204,1205,1212,1213,1231,1251,
1264,1268,1270,1284,1285,1311,1318,1321,1336,1337,1339,
1345,1355,1356,1358,1359,1362,1364,1367,1368,1369,1381,
1412,1417,1418,1437,1448,1449,1457,1491,1492,1493,1497,
1501,1526,1533,1534,1540,1549,1550,1554,1568,1569,1590,
1594,1595,1611,1618,1619,1624,1625,1626,1628,1639,1640,
1646,1647,1659,1673

## Periodic Multilayer Structure
1176,1180,1181,1182,1206,1209,1217,1222,1223,1224,1225,
1227,1228,1233,1237,1238,1239,1241,1244,1249,1250,1251,
1252,1253,1254,1257,1258,1259,1267,1272,1273,1280,1281,
1287,1288,1289,1290,1291,1303,1310,1319,1320,1322,1323,
1324,1326,1327,1330,1334,1335,1341,1342,1345,1346,1360,
1379,1380,1382,1383,1384,1388,1392,1393,1394,1396,1402,
1408,1409,1414,1415,1416,1422,1441,1464,1467,1468,1474,
1475,1476,1477,1478,1480,1481,1482,1487,1494,1502,1503,
1505,1507,1508,1517,1521,1522,1523,1524,1525,1529,1530,
1531,1535,1536,1537,1541,1542,1543,1544,1545,1548,1551,
1552,1555,1559,1560,1563,1564,1571,1572,1574,1575,1582,
1587,1592,1593,1601,1603,1613,1614,1627,1630,1635,1642,
1643,1648,1652,1654,1655,1657,1663,1665,1666,1671,1672

## Modulation Doping
1173,1174,1179,1183,1186,1192,1207,1216,1219,1220,1222,
1223,1226,1229,1230,1232,1234,1235,1247,1248,1256,1261,
1263,1265,1279,1286,1291,1292,1293,1295,1297,1298,1299,
1300,1304,1305,1306,1307,1308,1312,1313,1314,1315,1320,
1324,1329,1331,1332,1335,1347,1348,1349,1357,1360,1363,
1371,1375,1376,1377,1378,1385,1386,1392,1395,1398,1399,
1400,1401,1409,1411,1413,1419,1420,1421,1423,1424,1425,

1426,1427,1428,1429,1430,1431,1432,1435,1440,1441,1442,
1445,1446,1450,1451,1454,1455,1456,1460,1463,1469,1473,
1483,1484,1485,1486,1488,1489,1495,1496,1498,1505,1506,
1509,1512,1515,1522,1523,1527,1528,1532,1538,1539,1546,
1547,1548,1559,1578,1579,1580,1582,1583,1584,1585,1586,
1600,1606,1607,1615,1616,1617,1620,1621,1622,1623,1629,
1631,1636,1637,1644,1652,1660,1661,1662,1665,1667,1668,
1669,1670

## Electrical Properties
1173,1174,1175,1176,1179,1180,1181,1182,1183,1185,1186,
1187,1188,1189,1190,1191,1192,1194,1195,1199,1200,1201,
1202,1204,1205,1207,1210,1212,1213,1214,1215,1216,1217,
1218,1219,1220,1221,1224,1225,1226,1227,1228,1229,1230,
1232,1234,1235,1236,1237,1238,1239,1240,1241,1242,1243,
1244,1245,1246,1247,1248,1252,1254,1255,1256,1259,1260,
1261,1262,1263,1265,1267,1268,1269,1271,1273,1274,1275,
1276,1278,1280,1281,1286,1288,1289,1290,1291,1292,1293,
1294,1295,1296,1297,1298,1299,1300,1301,1302,1303,1304,
1305,1306,1308,1312,1314,1315,1316,1317,1319,1321,1322,
1323,1324,1326,1329,1331,1332,1335,1336,1344,1345,1346,
1347,1348,1349,1350,1351,1352,1353,1354,1355,1356,1357,
1360,1361,1364,1370,1371,1372,1373,1374,1375,1376,1377,
1378,1385,1386,1387,1389,1390,1392,1395,1397,1398,1399,
1400,1401,1403,1404,1405,1406,1409,1410,1411,1413,1414,
1419,1420,1423,1424,1426,1427,1428,1429,1430,1431,1432,
1433,1435,1437,1438,1439,1440,1441,1442,1443,1444,1445,
1446,1450,1451,1452,1453,1454,1455,1456,1458,1459,1460,
1462,1463,1469,1470,1471,1472,1473,1479,1483,1484,1485,
1486,1487,1488,1489,1490,1495,1496,1497,1498,1499,1500,
1506,1507,1508,1509,1510,1511,1512,1515,1516,1523,1526,
1527,1528,1529,1530,1531,1537,1538,1539,1540,1541,1542,
1544,1545,1546,1547,1548,1553,1555,1556,1557,1559,1561,
1563,1564,1565,1566,1570,1571,1572,1573,1575,1577,1578,
1579,1580,1581,1582,1583,1584,1585,1586,1587,1588,1589,
1590,1594,1595,1598,1599,1604,1605,1606,1607,1609,1613,
1615,1616,1617,1618,1620,1621,1622,1623,1627,1629,1631,
1632,1633,1634,1635,1636,1637,1638,1639,1640,1643,1644,
1648,1649,1650,1651,1653,1654,1655,1657,1658,1659,1660,
1661,1662,1665,1667,1668,1669,1670,1671

## Optical Properties
1176,1178,1181,1188,1171,1193,1199,1200,1201,1203,1208,
1209,1211,1216,1217,1222,1223,1224,1225,1233,1237,1238,
1239,1240,1241,1242,1243,1244,1252,1253,1257,1258,1259,
1260,1261,1262,1263,1266,1267,1268,1274,1275,1277,1280,
1282,1283,1287,1288,1289,1290,1291,1296,1313,1322,1325,
1327,1330,1333,1334,1335,1340,1341,1342,1343,1345,1346,
1363,1364,1365,1366,1379,1380,1382,1383,1384,1387,1388,
1389,1390,1391,1396,1402,1403,1406,1414,1415,1416,1422,
1425,1433,1439,1441,1447,1452,1461,1464,1466,1467,1468,
1470,1472,1474,1475,1476,1477,1478,1480,1481,1482,1488,
1490,1497,1500,1502,1504,1505,1507,1508,1512,1513,1514,
1517,1522,1523,1524,1525,1529,1531,1535,1536,1541,1543,
1544,1545,1551,1552,1558,1559,1560,1562,1563,1571,1574,

1577,1591,1592,1593,1596,1597,1601,1602,1603,1604,1608,
1609,1610,1611,1612,1613,1614,1630,1633,1634,1635,1638,
1640,1641,1642,1648,1652,1653,1654,1655,1656,1658,1659,
1663,1664,1666,1672

**Microwave Devices**
1173,1174,1175,1189,1190,1194,1195,1210,1214,1215,1216,
1218,1246,1260,1268,1278,1286,1292,1294,1295,1297,1300,
1312,1315,1344,1351,1352,1353,1354,1372,1374,1386,1395,
1404,1423,1424,1428,1431,1432,1441,1442,1446,1451,1462,
1473,1479,1483,1484,1485,1486,1500,1515,1539,1546,1553,
1556,1561,1587,1588,1589,1599,1605,1616,1621,1622,1623,
1632,1633,1634,1635,1655,1667,1668,1669,1670,1671

**Optoelectronic Devices**
1176,1178,1188,1191,1193,1216,1217,1237,1238,1239,1240,
1242,1243,1244,1260,1261,1262,1263,1266,1267,1268,1274,
1282,1283,1296,1325,1358,1365,1379,1380,1387,1390,1391,
1402,1406,1434,1441,1500,1502,1512,1531,1571,1592,1596,
1603,1609,1610,1611,1612,1632,1633,1634,1635,1638,1642,
1655,1656

1173  Abe,M., Mimura,T., Nishiuchi,K., Shibatomi,A., Kobayashi,M.
  R   HEMT LSI technology for high speed computers
      1983 GaAs IC Symposium Technical Digest    158-161
      (1983)

1174  Abe,M., Mimura,T., Yokoyama,N., Suyama,K.
  R   Advanced device technology for high speed GaAs VLSI
      in: Solid State Devices 1982/ESSDERC SSSDT Meeting,
         Munich, Sept. 13-16,1982; Eds. A. Goetzberger and
         M. Zerbst (Physik-Verlag, Weinheim, 1983)    25-50
         (1983)

1175  Aishima,A., Fukushima,Y.
  T   New negative conductances in GaAs $n(+)-n-n(+)$
         ballistic diodes
      Jpn. J. Appl. Phys. 22    L255-L257 (1983)

1176  Alavi,K., Pearsall,T.P., Forrest,S.R., Cho,A.Y.
      $Ga(0.47)In(0.53)As/Al(0.48)In(0.52)As$
         multiquantum-well LEDs emitting at 1.6 um
      Electron. Lett. 19    227-229 (1983)

1177  Alavi,K., Petroff,P.M., Wagner,W.R., Cho,A.Y.
      Substrate rotation-induced compositional oscillation
         in molecular beam epitaxy (MBE)
      J. Vac. Sci. Technol. B 1    146-148 (1983)

1178  Alavi,K., Temkin,H., Wagner,W.R., Cho,A.Y.
      Optically pumped 1.55-um double heterostructure
         $Ga(x)Al(y)In(1-x-y)As/Al(u)In(1-u)As$ lasers grown
         by molecular beam epitaxy
      Appl. Phys. Lett. 42    254-256 (1983)

1179  Allen,S.J.,Jr., Stoermer,H.L., Hwang,J.C.M.
      Dimensional resonance of the two-dimensional electron
         gas in selectively doped GaAs/AlGaAs
         heterostructures
      Phys. Rev. B 28    4875-4877 (1983)

1180  Altarelli,M.
  T   Electronic structure and semiconductor-semimetal
         transition in InAs-GaSb superlattices
      Phys. Rev. B 28    842-845 (1983)

1181  Altarelli,M.
  TR  Electronic structure of semiconductor superlattices
      Lect. Notes Phys. 177    174-185 (1983)

1182  Altarelli,M.
  T   Electronic structure of semiconductor superlattices
      Physica 117B & 118B    747-749 (1983)

1183    **Anderson,P.W.**
T       Remarks on the Laughlin theory of the fractionally
            quantized Hall effect
        Phys. Rev. B 28    2264-2265 (1983)

1184    **Andersson,T.G., Nilsson,B., Svensson,S.P.,
        Flemming,K.E.**
        A combined vacuum interlock and preparation assembly
            for MBE and surface analysis
        J. Phys. E 16    364-366 (1983)

1185    **Andersson,T.G., Svensson,S.P., Landgren,G.**
        Initial growth of Al on GaAs(001) and electrical
            characterization of the interface
        J. Vac. Sci. Technol. B 1    361-364 (1983)

1186    **Ando,T.**
T       Electron localization in a two-dimensional system in
            strong magnetic fields. I.) Case of short-range
            scatterers
        J. Phys. Soc. Jpn. 52    1740-1749 (1983)

1187    **Andrews,D.A., Heckingbottom,R., Davies,G.J.**
        The influence of growth conditions on sulfur
            incorporation in GaAs grown by molecular beam
            epitaxy
        J. Appl. Phys. 54    4421-4425 (1983)

1188    **Ankri,D., Schaff,W.J., Barnard,J.A., Lunardi,L.,
        Eastman,L.F.**
        High-speed GaAs heterojunction bipolar
            phototransistor grown by molecular beam epitaxy
        Electron. Lett. 19    278-280 (1983)

1189    **Ankri,D., Schaff,W.J., Smith,P., Eastman,L.F.**
        High-speed GaAlAs-GaAs heterojunction bipolar
            transistors with near-ballistic operation
        Electron. Lett. 19    147-149 (1983)

1190    **Ankri,D., Schaff,W.J., Wood,C.E.C., Eastman,L.F.,
        Woodard,D.W., Rathbun,L.**
        GaAlAs/GaAs heterojunction bipolar transistors with
            abrupt emitter base interface for ballistic
            operation
        Inst. Phys. Conf. Ser. 65    431-438 (1983)

1191    **Anthony,P.J., Pawlik,J.R., Swaminathan,V.,
        Tsang,W.T.**
        Reduced threshold current temperature dependence in
            double heterostructure lasers due to separate p-n
            and heterojunctions
        IEEE J. Quantum Electron. QE-19    1030-1035 (1983)

1192 Aoki,H.
TR  Gauge invariance and the quantised Hall effect in
    two-dimensional systems
    Lect. Notes Phys. 177   11-22 (1983)

1193 Asahi,H., Kawamura,Y., Noguchi,Y., Matsuoka,T.,
       Nagai,H.
     Hybrid LPE/MBE-grown InGaAsP/InP DFB lasers
     Electron. Lett. 19   507-509 (1983)

1194 Asbeck,P.M., Miller,D.L., Anderson,R.J.,
       Eisen,F.H.
     Emitter-coupled logic circuits implemented with
       heterojunction bipolar transistors
     1983 GaAs IC Symposium Technical Digest   170-173
       (1983)

1195 Asbeck,P.M., Miller,D.L., Babcock,E.J.,
       Kirkpatrick,C.G.
     Application of thermal pulse annealing to
       ion-implanted GaAsAs/GaAs heterojunction bipolar
       transistors
     IEEE Electron Device Lett. EDL-4   81-83 (1983)

1196 Bachrach,R.Z., Bringans,R.D.
     On the possibility of MBE growth interface
       modification by hydrogen
     J. Vac. Sci. Technol. B 1   142-145 (1983)

1197 Bachrach,R.Z., Bringans,R.D.
     On the possibility of MBE growth interface
       modification by hydrogen
     J. Vac. Sci. Technol. B 1   142-145 (1983)

1198 Bafleur,M., Munoz-Yague,A.
     Crystal, impurity-related and growth defects in
       molecular beam epitaxial GaAs layers
     Thin Solid Films 101   299-310 (1983)

1199 Bafleur,M., Munoz-Yague,A., Castano,J.L.,
       Piqueras,J.
     Photoluminescence of molecular beam epitaxially grown
       Ge-doped GaAs
     J. Appl. Phys. 54   2630-2634 (1983)

1200 Ballingall,J.M., Collins,D.M.
     Photoluminescence of shallow acceptors in
       Al(0.28)Ga(0.72)As
     J. Appl. Phys. 54   341-345 (1983)

1201 Ballingall,J.M., Wood,C.E.C.
     Autocompensation in molecular beam epitaxial gallium
       arsenide: The (110) orientation
     J. Vac. Sci. Technol. B 1   162-165 (1983)

1202 Ballingall,J.M., Wood,C.E.C., Eastman,L.F.
Electrical measurements of the conduction band discontinuity of the abrupt Ge-GaAs<100> heterojunction
J. Vac. Sci. Technol. B 1    675-681 (1983)

1203 Bamba,Y., Miyauchi,E., Kuramoto,K., Takamori,A., Furuya,T.
Sn ion doping during GaAs MBE with field ion gun
Jpn. J. Appl. Phys. 22    L331-L332 (1983)

1204 Barret,C., Chekir,F., Nefatti,T., Vapaille,A., Massies,J.
Stoichiometry effects on Schottky barrier and interface states in GaAs(001)/Al system
Physica 117B & 118B    851-853 (1983)

1205 Barret,C., Massies,J.
On the dependence of Schottky barrier heights and interface states upon initial semiconductor surface parameters in GaAs(001)/Al junctions
J. Vac. Sci. Technol. B 1    819-824 (1983)

1206 Barrett,J.H.
T    Ion dechanneling due to lattice strains in semiconductor superlattices
Phys. Rev. B 28    2328-2334 (1983)

1207 Bastard,G.
T    Energy levels and alloy scattering in InP-In(Ga)As heterojunctions
Appl. Phys. Lett. 43    591-593 (1983)

1208 Bastard,G., Mendez,E.E., Chang,L.L., Esaki,L.
T    Far infrared impurity absorption in a quantum well
Solid State Commun. 45    367-369 (1983)

1209 Bastard,G., Mendez,E.E., Chang,L.L., Esaki,L.
T    Variational calculations on a quantum well in an electric field
Phys. Rev. B 28    3241-3245 (1983)

1210 Basu,P.K., Bhattacharyya,K., Nag,B.R.
T    Mobility of two-dimensional electron gas in JFETs limited by polar-optic and impurity scattering
Solid State Commun. 48    981-984 (1983)

1211 Baudet,M., Regreny,O., Dupas,G., Auvray,P., Gauneau,M., Regreny,A., Talalaeff,G.
Dosage de l'aluminium par spectrometrie d'absorption atomique et diffraction des rayons X dans des couches epitaxiees par jets moleculaires de $Ga(1-x)Al(x)As$
Mat. Res. Bull. 18    123-133 (1983)

1212   Bauer,R.S., Sang,H.W.,Jr.
   R   On the adjustability of the "abrupt" heterojunction
       band-gap discontintuity
       Surf. Sci. 132   479-504 (1983)

1213   Bauer,R.S., Zurcher,P., Sang,H.W.,Jr.
       Inequality of semiconductor heterojunction
       conduction-band-edge discontinuity and electron
       affinity difference
       Appl. Phys. Lett. 43   663-665 (1983)

1214   Bayraktaroglu,B., Shih,H.D.
       Integral packaging for millimetre-wave GaAs impatt
       diodes prepared by molecular beam epitaxy
       Electron. Lett. 19   327-329 (1983)

1215   Bayraktaroglu,B., Shih,H.D.
       Millimeter-wave GaAs distributed IMPATT diodes
       IEEE Electron Device Lett. EDL-4   393-395 (1983)

1216   Bethea,C.G., Chen,C.Y., Cho,A.Y., Garbinski,P.A.,
       Levine,B.F.
       Picosecond Al(x)Ga(1-x)As modulation-doped optical
       field-effect transistor sampling gate
       Appl. Phys. Lett. 42   682-684 (1983)

1217   Bhattacharya,B.K., Tripathi,V.K.
   T   Lattice-matched and mismatched multiquantum-well
       heterostructure photodiodes for operation at 1.1
       to 1.5 um
       Electron. Lett. 19   924-926 (1983)

1218   Blakey,P.A., East,J.R., Elta,M.E., Haddad,G.I.
   T   Implications of velocity overshoot in heterojunction
       transit-time diodes
       Electron. Lett. 19   510-512 (1983)

1219   Bliek,L., Braun,E., Engelmann,H.J., Leontiew,H.,
       Melchert,F., Schlapp,W., Stahl,B., Warnecke,P.,
       Weimann,G.
       Measurements of the quantized Hall resistance
       $h/(2*e**2)$ with a reproducibility of $10**(-8)$ and
       its application for a novel determination of the
       Ohm
       PTB -Mitt. 93   21-23 (1983)

1220   Bliek,L., Braun,E., Melchert,F., Warnecke,P.,
       Schlapp,W., Weimann,G., Ploog,K., Ebert,G.
       Praezisionsmessungen des quantisierten
       Hall-Widerstandes und Bestimmung der
       Feinstrukturkonstanten
       Phys. Bl. 39   157-158 (1983)

1221  Blood,P., Grassie,A.D.C.
      Optical excitation of defects in molecular beam
        epitaxy grown GaAs with polarized light
      Phys. Rev. B 27   2548-2550 (1983)

1222  Bloss,W.L.
    T Collective plasmon modes of a semiconductor
        superlattice
      J. Vac. Sci. Technol. B 1   431-434 (1983)

1223  Bloss,W.L.
    T Intersubband plasmons of a semiconductor superlattice
      Solid State Commun. 46   143-146 (1983)

1224  Bloss,W.L.
    T Plasmon modes of a superlattice - classical vs
        quantum limits
      Solid State Commun. 48   927-931 (1983)

1225  Bloss,W.L., Friedman,L.
    T Enhanced optical nonlinearities in superlattices
      J. Vac. Sci. Technol. B 1   150-151 (1983)

1226  Bok,J., Combescot,M.
    T Capacitance oscillations in quantized Hall effect
      Solid State Commun. 47   611-613 (1983)

1227  Borovitskaya,E.S., Genkin,V.M.
    T To the theory of I-V characteristics of the
        superlattices
      Solid State Commun. 46   769-771 (1983)

1228  Brand,S., Abram,R.A.
    T Self-consistent calculations of electron and hole
        subband energies for an n-p superlattice in GaAs
      J. Phys. C 16   6111-6120 (1983)

1229  Braun,E., Kose,V., Melchert,F., Warnecke,P.
    T Possible application of the quantized Hall resistance
        to the realization of the electrical units
      Lect. Notes Phys. 177   37-40 (1983)

1230  Brenig,W.
    T Quantized Hall conductance: effect of random
        potentials
      Z. Phys. B 50   305-309 (1983)

1231  Bringans,R.D., Bachrach,R.Z.
      Hydrogen chemisorption on the polar surfaces of GaAs
      J. Vac. Sci. Technol. A 1   676-678 (1983)

1232  Brummell,M.A., Nicholas,R.J., Portal,J.C.,
        Cheng,K.Y., Cho,A.Y.
      Two-dimensional magnetophonon resonance: II.
        GaInAs-AlInAs heterojunctions
      J. Phys. C 16   L579-L584 (1983)

1233 Busserand,B., Paquet,D., Regreny,A., Kervarec,J.
Raman scattering determination of folded acoustical phonon dispersion curves in large period GaAs/GaAlAs superlattices
Solid State Commun. 48   499-502 (1983)

1234 Cage,M.E., Dziuba,R.F., Field,B.F., Williams,E.R., Girvin,S.M., Gossard,A.C., Tsui,D.C., Wagner,R.J.
Dissipation and dynamic nonlinear behavior in the quantum Hall regime
Phys. Rev. Lett. 51   1374-1377 (1983)

1235 Cage,M.E., Girvin,S.M.
R The quantum Hall effect. I
Comments Solid State Phys. 11   1-16 (1983)

1236 Campagna,M., Alvarado,S.F., Ciccacci,F.
Solid state polarized electron sources
AIP Conf. Proc. 95   566-573 (1983)

1237 Capasso,F.
R Band-gap engineering via graded gap, superlattice, and periodic doping structures: application to novel photodetectors and other devices
J. Vac. Sci. Technol. B 1   457-461 (1983)

1238 Capasso,F.
R New device applications of bandedge discontinuities in multilayer heterojunction structures
Surf. Sci. 132   527-539 (1983)

1239 Capasso,F.
R Avalanche photodiodes with enhanced ionization rates ratio: towards a solid state photomultiplier
IEEE Trans. Nucl. Sci. NS-30   424-428 (1983)

1240 Capasso,F., Alavi,K., Cho,A.Y., Foy,P.W., Bethea,C.G.
Long wavelength, wide spectral response (0.8-1.8 um) Al(0.48)In(0.52)As/Ga(0.47)In(0.53)As avalanche photodiodes and Al(0.48)In(0.52)As electroabsorption pin avalanche detectors grown by molecular beam epitaxy
IEDM 83 Technical Digest   468-471 (1983)

1241 Capasso,F., Luryi,S., Tsang,W.T., Bethea,C.G., Levine,B.F.
New transient electrical polarization phenomenon in sawtooth superlattices
Phys. Rev. Lett. 51   2318-2321 (1983)

1242 Capasso,F., Tsang,W.T., Bethea,C.G., Hutchinson,A.L., Levine,B.F.
New graded band-gap picosecond phototransistor
Appl. Phys. Lett. 42   93-95 (1983)

1243  Capasso,F., Tsang,W.T., Hutchinson,A.L.,
      Bethea,C.G., Levine,B.F.
      Novel graded band-gap photodetectors with ultrahigh
        speed of response (tr = 20 ps, FWHM = 40 ps) and
        phototransistors with graded gap base
      Inst. Phys. Conf. Ser. 65   225-231 (1983)

1244  Capasso,F., Tsang,W.T., Williams,G.F.
   R  Staircase solid-state photomultipliers and avalanche
        photodiodes with enhanced ionization rates ratio
      IEEE Trans. Electron Devices ED-30   381-390 (1983)

1245  Carlsson,A.E., Ehrenreich,H., Hass,K.C.
   T  Spectral limits for disordered semiconductors and
        their interfaces
      Phys. Rev. B 28   4468-4471 (1983)

1246  Chai,Y.G., Yeats,R.
      In(0.53)Ga(0.47)As submicrometer FET's grown by MBE
      IEEE Electron Device Lett. EDL-4   252-254 (1983)

1247  Chalker,J.T.
   T  The Hall effect in a two-dimensional electron gas
      J. Phys. C 16   4597-4304 (1983)

1248  Chang,A.M., Paalanen,M.A., Tsui,D.C.,
      Stoermer,H.L., Hwang,J.C.M.
      Fractional quantum Hall effect at low temperatures
      Phys. Rev. B 28   6133-6136 (1983)

1249  Chang,C.A.
      Channeling studies of III-V/III-V and IV/III-V
        semiconductor modulated structures
      J. Vac. Sci. Technol. B 1   346-352 (1983)

1250  Chang,C.A., Chu,W.K.
      Channeling studies of Ge-GaAs superlattices grown by
        molecular beam epitaxy
      Appl. Phys. Lett. 42   463-465 (1983)

1251  Chang,C.A., Kuan,T.S.
      Structural studies of Ge-GaAs interfaces
      J. Vac. Sci. Technol. B 1   315-319 (1983)

1252  Chang,L.L.
   R  A review of recent advances in semiconductor
        superlattices
      J. Vac. Sci. Technol. B 1   120-125 (1983)

1253  Chang,Y.C., Schulman,J.N.
   T  Modification of optical properties of
        GaAs-Ga(1-x)Al(x)As superlattices due to band
        mixing
      Appl. Phys. Lett. 43   536-538 (1983)

1254 **Chang,Y.C., Ting,D.Z.Y.**
T    Interference effect in multivalley quantum well
       structures
     J. Vac. Sci. Technol. B 1   435-438 (1983)

1255 **Chattopadhyay,D.**
T    Transient behaviour of hot electrons in quantum-well
       structures of polar semiconductors
     Phys. Status Solidi B 119   K77-K81 (1983)

1256 **Chattopadhyay,D., Ghosal,A.**
T    Two-dimensional hot-electron transport in
       quantum-well structures of polar semiconductors
     J. Phys. C 16   2583-2586 (1983)

1257 **Chaudhuri,S.**
T    Hydrogenic-impurity ground state in
       GaAs-Ga(1-x)Al(x)As multiple-quantum-well
       structures
     Phys. Rev. B 28   4480-4488 (1983)

1258 **Chemla,D.S.**
R    Quasi-two-dimensional excitons in GaAs/Al(x)Ga(1-x)As
       semiconductor multiple quantum well structures
     Helv. Phys. Acta 56   607-637 (1983)

1259 **Chemla,D.S., Damen,T.C., Miller,D.A.B.,
       Gossard,A.C., Wiegmann,W.**
     Electroabsorption by Stark effect on room-temperature
       excitons in GaAs/GaAlAs multiple quantum well
       structures
     Appl. Phys. Lett. 42   864-866 (1983)

1260 **Chen,C.Y., Alavi,K., Cho,A.Y., Garbinski,P.A.,
       Pang,Y.M.**
     New Ga(0.47)In(0.53)As sheet-charge field-effect
       transistor for long-wavelength optoelectronic
       integration
     Electron. Lett. 19   791-792 (1983)

1261 **Chen,C.Y., Cho,A.Y., Bethea,C.G., Garbinski,P.A.,
       Pang,Y.M., Levine,B.F., Ogawa,K.**
     Ultrahigh speed modulation-doped heterostructure
       field-effect photodetectors
     Appl. Phys. Lett. 42   1040-1042 (1983)

1262 **Chen,C.Y., Pang,Y.M., Cho,A.Y., Garbinski,P.A.**
     Removing long tails from photoconductive detectors: a
       new minority hole sinked photodetector
     IEDM 83 Technical Digest   643-646 (1983)

1263  Chen,C.Y., Pang,Y.M., Garbinski,P.A., Cho,A.Y., Alavi,K.
Modulation-doped Ga(0.47)In(0.53)As/Al(0.48)In(0.52)As planar photoconductive detectors for 1.0-1.55-um applications
Appl. Phys. Lett. 43    308-310 (1983)

1264  Chiang,T.C., Ludeke,R., Aono,M., Landgren,G., Himpsel,F.J., Eastman,D.E.
Angle-resolved photoemission studies of GaAs(100) surfaces grown by molecular-beam epitaxy
Phys. Rev. B 27    4770-4778 (1983)

1265  Chiu,L.C., Margalit,S., Yariv,A.
T  Intrinsic limitations on the high speed performance of high electron mobility transistors
Jpn. J. Appl. Phys. 22    L82-L84 (1983)

1266  Chiu,L.C., Smith,J.S., Margalit,S., Yariv,A.
T  Internal photoemission from quantum well heterojunction superlattices by phononless free-carrier absorption
Appl. Phys. Lett. 43    331-332 (1983)

1267  Chiu,L.C., Smith,J.S., Margalit,S., Yariv,A., Cho,A.Y.
Application of internal photoemission from quantum-well and heterojunction superlattices to infrared photodetectors
Infrared Phys. 23    93-97 (1983)

1268  Cho,A.Y.
R  Growth of III-V semiconductors by molecular beam epitaxy and their properties
Thin Solid Films 100    291-317 (1983)

1269  Choudhury,A.N.M.M., Rowe,W., Tabatabaie-Alavi,K., Fonstad,C.G., Alavi,K., Cho,A.Y.
Ion implantation of Si And Be in Al(0.48)In(0.52)As
J. Appl. Phys. 54    4374-4377 (1983)

1270  Chow,R., Chai,Y.G.
A PH3 cracking furnace for molecular beam epitaxy
J. Vac. Sci. Technol. A1    49-54 (1983)

1271  Chow,R., Chai,Y.G.
Electrical and optical properties of InP grown by molecular beam epitaxy using cracked phosphine
Appl. Phys. Lett. 42    383-385 (1983)

1272  Chu,W.K., Pan,C.K., Chang,C.A.
Superlattice interface and lattice strain measurement by ion channeling
Phys. Rev. B 27    4033-4036 (1983)

1273  Churchill,J.N., Holmstrom,F.E.
T     Energy states and Bloch states for an accelerated
         electron in a periodic lattice
      Phys. Scr. 27   91-98 (1983)

1274  Collins,D.M., Mars,D.E., Eglash,S.J.
      The growth of high quality Al(x)Ga(1-x)As by
         molecular beam epitaxy and its application to
         double-heterojunction lasers
      J. Vac. Sci. Technol. B 1   170-173 (1983)

1275  Collins,D.M., Mars,D.E., Fischer,B., Kocot,C.
      Heterojunction-induced phenomena in Hall effect and
         photoconductivity measurements of epitaxial
         Al(x)Ga(1-x)As
      J. Appl. Phys. 54   857-861 (1983)

1276  Collins,D.M., Mars,D.E., Fischer,B., Kocot,C.
      Heterojunction-induced phenomena in Hall effect and
         photoconductivity measurements of epitaxial
         Al(x)Ga(1-x)As
      Inst. Phys. Conf. Ser. 65   581-588 (1983)

1277  Contour,J.P., Neu,G., Leroux,M., Chaix,C.,
         Levesque,B., Etienne,P.
      An optical characterization of defect levels induced
         by MBE growth of GaAs
      J. Vac. Sci. Technol. B 1   811 (1983)

1278  Cook,R.K.
T     Computer simulation of carrier transport in planar
         doped barrier diodes
      Appl. Phys. Lett. 42   439-441 (1983)

1279  Das Sarma,S.
T     Polaron effective mass in GaAs heterostructure
      Phys. Rev. B 27   2590-2593 (1983)

1280  Das Sarma,S.
T     Dispersion of magnetoplasmons in layered systems
      Phys. Rev. B 28   2240-2243 (1983)

1281  Datta,S., Gunshor,R.L.
T     Space-charge waves in multilayered heterostructures
      J. Appl. Phys. 54   4453-4456 (1983)

1282  Dawson,M.D., Sibbett,W., Vukusic,J.I., Dawson,P.,
         Duggan,G., Foxon,C.T.
      Streak camera study of short pulse generation in an
         optically pumped GaAs/(GaAl)As laser
      Appl. Phys. Lett. 43   226-228 (1983)

1283 DeFreez,R.K., Elliott,R.A., Blakemore,J.S.,
Miller,B.I., McFee,J.H., Martin,R.J.
High-output room-temperature pulsed operation for
broad contact InP/In(0.53)Ga(0.47)As/InP lasers
grown by molecular beam epitaxy
J. Appl. Phys. 54   2177-2182 (1983)

1284 DeJong,T., Douma,W.A.S., Veen,J.F. van der,
Saris,F.W., Haisma,J.
Silicon molecular beam epitaxy on gallium phosphide
Appl. Phys. Lett. 42   1037-1039 (1983)

1285 DeJong,T., Douma,W.A.S., Veen,J.F. van der,
Saris,F.W., Haisma,J.
Silicon molecular beam epitaxy on gallium phosphide
Appl. Phys. Lett. 42   1037-1039 (1983)

1286 Delagebeaudeuf,D., Delescluse,P., Laviron,M.,
Tung,P.N., Chaplart,J., Chevrier,J., Linh,N.T.
Low and high field transport properties in
two-dimensional electron gas FET
Inst. Phys. Conf. Ser. 65   393-398 (1983)

1287 Djafari-Rouhani,B., Dobrzynski,L.,
Hardouin-Duparc,O.
T  Surface phonons in superlattices
J. Electron Spectrosc. Relat. Phenom. 30   119-124
(1983)

1288 Doehler,G.H.
TR  n-i-p-i doping superlattices - metastable
semiconductors with tunable properties
J. Vac. Sci. Technol. B 1   278-284 (1983)

1289 Doehler,G.H.
TR  n-i-p-i superlattices - novel semiconductors with
tunable properties
Jpn. J. Appl. Phys. 22, Suppl. 22-1   29-35 (1983)

1290 Doehler,G.H.
R  n-i-p-i doping superlattices - tailored
semiconductors with tunable electronic properties
in "Festkoerperprobleme", Ed. P. Grosse (Vieweg,
Braunschweig, 1983) Vol. XXIII   207-226 (1983)

1291 Doehler,G.H.
R  Solid-state superlattices
Scientific American 249 No. 5   118-126 (1983)

1292 Drummond,T.J., Fischer,R., Su,S.L., Lyons,W.G.,
Morkoc,H., Lee,K., Shur,M.S.
Characteristics of modulation-doped
Al(x)Ga(1-x)As/GaAs field-effect transistors:
effect of donor-electron separation
Appl. Phys. Lett. 42   262-264 (1983)

1293 Drummond,T.J., Klem,J., Arnold,D., Fischer,R.,
     Thorne,R.E., Lyons,W.G., Morkoc,H.
     Use of a superlattice to enhance the interface
         properties between two bulk heterolayers
     Appl. Phys. Lett. 42    615-617 (1983)

1294 Drummond,T.J., Kopp,W., Arnold,D., Fischer,R.,
     Morkoc,H., Erickson,L.P., Palmberg,P.W.
     Enhancement-mode metal/(Al,Ga)As/GaAs
         buried-interface field-effect transistor (BIFET)
     Electron. Lett. 19    986-988 (1983)

1295 Duh,K.H., Ziel,A. van der, Morkoc,H.
     1/f noise in modulation-doped field effect transistors
     IEEE Electron Device Lett. EDL-4    12-13 (1983)

1296 Dutta,N.K., Hartman,R.L., Tsang,W.T.
     Gain and carrier lifetime measurements in AlGaAs
         single quantum well lasers
     IEEE J. Quantum Electron. QE-19    1243-1246 (1983)

1297 Eastman,L.F.
   R Use of molecular beam epitaxy in research and
         development of selected high speed compound
         semiconductor devices
     J. Vac. Sci. Technol. B 1    131-134 (1983)

1298 Ebert,G., Klitzing,K. von, Ploog,K., Weimann,G.
     Two-dimensional magneto-quantum transport on
         GaAs-Al(x)Ga(1-x)As heterostructures under
         non-ohmic conditions
     J. Phys. C 16    5441-5448 (1983)

1299 Ebert,G., Klitzing,K. von, Probst,C., Schuberth,E.,
     Ploog,K., Weimann,G.
     Hopping conduction in the Landau level tails in
         GaAs-Al(x)Ga(1-x)As heterostructures at low
         temperatures
     Solid State Commun. 45    625-628 (1983)

1300 Eden,R.C., Livingston,A.R., Welch,B.M.
   R Integrated circuits: the case for gallium arsenide
     IEEE Spectrum 20 No. 12    30-37 (1983)

1301 Eglash,S.J., Newman,N., Pan,S., Spicer,W.E.,
     Collins,D.M., Zurakowski,M.P.
     Barrier heights from the ohmic to bandgap: modified
         Al:GaAs Schottky diodes by MBE
     IEDM 83 Technical Digest    119-122 (1983)

1302 Eglash,S.J., Pan,S., Mo,D., Spicer,W.E.,
     Collins,D.M.
     Modified Schottky barrier heights by interfacial
         doped layers: MBE Al on GaAs
     Jpn. J. Appl. Phys. 22, Suppl. 22-1    431-435 (1983)

1303 **Ekenberg,U., Hess,K.**
T  Impurity-band tails in superlattices
   Phys. Rev. B 27   3445-3450 (1983)

1304 **Englert,T.**
R  High field magnetotransport in GaAs/AlGaAs
      heterojunctions and Si MOSFETs
   Lect. Notes Phys. 177   87-97 (1983)

1305 **Englert,T., Maan,J.C., Tsui,D.C., Gossard,A.C.**
   A study of intersubband scattering in
      GaAs/Al(x)Ga(1-x)As heterostructures by means of
      a parallel magnetic field
   Solid State Commun. 45   989-991 (1983)

1306 **Englert,T., Maan,J.C., Uihlein,C., Tsui,D.C., Gossard,A.C.**
   Cyclotron resonance of 2D electrons in
      GaAs/Al(x)Ga(1-x)As heterostructures at low
      densities
   J. Vac. Sci. Technol. B 1   427-430 (1983)

1307 **Englert,T., Maan,J.C., Uihlein,C., Tsui,D.C., Gossard,A.C.**
   Observation of oscillatory linewidth in the cyclotron
      resonance of GaAs-Al(x)Ga(1-x)As heterostructures
   Solid State Commun. 46   545-548 (1983)

1308 **Englert,T., Maan,J.C., Uihlein,C., Tsui,D.C., Gossard,A.C.**
   Oscillations of the cyclotron resonance linewidth
      with Landau level filling factor in GaAs/AlGaAs
      heterostructures
   Physica 117B & 118B   631-633 (1983)

1309 **Erickson,L.P., Mattord,T.J., Palmberg,P.W., Fischer,R., Morkoc,H.**
   Growth of Al(0.3)Ga(0.7)As by molecular beam epitaxy
      in the forbidden temperature range using As2
   Electron. Lett. 19   632-633 (1983)

1310 **Erickson,L.P., Phillips,B.F.**
   Examination of MBE GaAs/Al(0.3)Ga(0.7)As
      superlattices by Auger electron spectroscopy
   J. Vac. Sci. Technol. B 1   158-161 (1983)

1311 **Farrow,R.F.C.**
R  The stabilization of metastable phases by epitaxy
   J. Vac. Sci. Technol. B 1   222-228 (1983)

1312 **Feuer,M.D., Hendel,R.H., Kiehl,R.A., Hwang,J.C.M., Keramidas,V.G., Allyn,C.L., Dingle,R.**
   High-speed low-voltage ring oscillators based on
      selectively doped heterojunction transistors
   IEEE Electron Device Lett. EDL-4   306-307 (1983)

1313   Fischer,B., Collins,D.M.
       Persistent red shift of the photoluminescence from
          the semi-insulating GaAs substrate in an
          Al(x)Ga(1-x)As/GaAs modulation-doped structure
       J. Vac. Sci. Technol. B 1   420-422 (1983)

1314   Fischer,R., Arnold,D., Thorne,R.E., Drummond,T.J.,
          Morkoc,H.
       Light sensitivity of Al(0.25)Ga(0.75)As/GaAs
          modulation-doped structures grown by molecular
          beam epitaxy: effect of substrate temperature
       Electron. Lett. 19   200-202 (1983)

1315   Fischer,R., Drummond,T.J., Kopp,W., Morkoc,H.,
          Lee,K., Shur,M.S.
       Instabilities in modulation doped field-effect
          transistors (MODFETs) at 77 K
       Electron. Lett. 19   789-791 (1983)

1316   Fischer,R., Drummond,T.J., Thorne,R.E., Lyons,W.G.,
          Morkoc,H.
       Properties of silicon-doped Al(x)Ga(1-x)As grown by
          molecular beam epitaxy
       Thin Solid Films 99   391-397 (1983)

1317   Fischer,R., Hopkins,C.G., Evans,C.A.,Jr.,
          Drummond,T.J., Lyons,W.G., Klem,J., Colvard,C.,
          Morkoc,H.
       The properties of Si in MBE grown Al(x)Ga(1-x)As
       Inst. Phys. Conf. Ser. 65   157-164 (1983)

1318   Fischer,R., Klem,J., Drummond,T.J., Thorne,R.E.,
          Kopp,W., Morkoc,H., Cho,A.Y.
       Incorporation rates of gallium and aluminum on GaAs
          during molecular beam epitaxy at high substrate
          temperatures
       J. Appl. Phys. 54   2508-2510 (1983)

1319   Fishman,G.
  T    Energy levels and Coulomb matrix elements in doped
          GaAs-(GaAl)As multiple-quantum-well
          heterostructures
       Phys. Rev. B 27   7611-7623 (1983)

1320   Fishman,G., Calecki,D.
  T    Mobility in modulation doped GaAs-GaAlAs superlattices
       Physica 117B & 118B   744-746 (1983)

1321   Foxon,C.T.
  R    MBD growth of GaAs and III-V alloys
       J. Vac. Sci. Technol. B 1   293-297 (1983)

1322 Fritz,I.J., Dawson,L.R., Osbourn,G.C.,
     Gourley,P.L., Biefeld,R.M.
     MBE growth and characterization of InGaAs/GaAs
        strained-layer superlattices
     Inst. Phys. Conf. Ser. 65   241-247 (1983)

1323 Fritz,I.J., Dawson,L.R., Zipperian,T.E.
     Doping and transport studies in In(x)Ga(1-x)As/GaAs
        strained-layer superlattices
     J. Vac. Sci. Technol. B 1   387-390 (1983)

1324 Fritz,I.J., Dawson,L.R., Zipperian,T.E.
     Electron mobilities in In(0.2)Ga(0.8)As/GaAs
        strained-layer superlattices
     Appl. Phys. Lett. 43   846-848 (1983)

1325 Fujii,T., Hiyamizu,S., Wada,O., Sugahara,T.,
     Yamakoshi,S., Sakurai,T., Hashimoto,H.
     Extremely uniform GaAs-AlGaAs heterostructure layers
        with high optical quality by molecular beam epitaxy
     J. Cryst. Growth 61   393-396 (1983)

1326 Ghatak,K.P., Chakravarti,A.N.
   T Carrier statistics in superlattices of small-gap
        semiconductors in the presence of a quantizing
        magnetic field
     Phys. Status Solidi B 117   707-715 (1983)

1327 Gibbs,H.M., Jewell,J.L., Moloney,J.V., Tai,K.,
     Tarng,S.S., Weinberger,D.A., Gossard,A.C.,
     McCall,S.L., Passner,A., Wiegmann,W.
   R Optical bistability, regenerative pulsations, and
        transverse effects in room-temperature GaAs-AlGaAs
        superlattice etalons
     J. Physique 44 Colloque C2   C2/195-C2/204 (1983)

1328 Gibson,J.M., Phillips,J.M.
     Analysis of epitaxial fluoride-semiconductor
        interfaces
     Appl. Phys. Lett. 43   828-830 (1983)

1329 Girvin,S.M., Cage,M.E.
  TR The quantum Hall effect  II
     Comments Solid State Phys. 11   47-58 (1983)

1330 Giuliani,G.F., Quinn,J.J.
   T Charge-density excitations at the surface of a
        semiconductor superlattice: a new type of surface
        polariton
     Phys. Rev. Lett. 51   919-922 (1983)

1331 Giuliani,G.F., Quinn,J.J., Ying,S.C.
   T Quantization of the Hall conductance in a
        two-dimensional electron gas
     Phys. Rev. B 28   2969-2978 (1983)

1332 Glasser,M.L.
T    Longitudinal dielectric behavior of a two-dimensional
        electron gas in a uniform magnetic field
     Phys. Rev. B 28    4387-4396 (1983)

1333 Goebel,E.O., Jung,H., Kuhl,J., Ploog,K.
     Recombination enhancement due to carrier localization
        in quantum well structures
     Phys. Rev. Lett. 51    1588-1591 (1983)

1334 Gonzales de la Cruz,G., Tselis,A., Quinn,J.J.
T    Collective modes in semiconductor superlattices
     J. Phys. Chem. Solids 44    807-812 (1983)

1335 Gossard,A.C.
R    Quantum effects at GaAs/Al(x)Ga(1-x)As junctions
     Thin Solid Films 104    279-284 (1983)

1336 Gotoh,H., Sasamoto,K., Kuroda,S., Kimata,M.
     Molecular beam epitaxy of AlSb on GaAs and GaSb on
        AlSb films
     Phys. Status Solidi A 75    641-645 (1983)

1337 Gourrier,S., Smit,L., Friedel,P., Larsen,P.K.
     Photoemission studies of molecular beam epitaxially
        grown GaAs(001) surfaces exposed to a nitrogen
        plasma
     J. Appl. Phys. 54    3993-3997 (1983)

1338 Gowers,J.P.
     TEM image contrast from clustering in Ga-In
        containing III-V alloys
     Appl. Phys. A 31    23-27 (1983)

1339 Grant,R.W., Kraut,E.A., Kowalczyk,S.P.,
        Waldrop,J.R.
R    Measurement of potential at semiconductor interfaces
        by electron spectroscopy
     J. Vac. Sci. Technol. B 1    320-327 (1983)

1340 Greene,J.E.
R    A review of recent research on the growth and
        physical properties of single crystal metastable
        elemental and alloy semiconductors
     J. Vac. Sci. Technol. B 1    229-237 (1983)

1341 Greene,R.L., Bajaj,K.K.
T    Energy levels of hydrogenic impurity states in
        GaAs-Ga(1-x)Al(x)As quantum well structures
     Solid State Commun. 45    825-829 (1983)

1342 Greene,R.L., Bajaj,K.K.
T    Binding energies of Wannier excitons in
        GaAs-Ga(1-x)Al(x)As quantum well structures
     Solid State Commun. 45    831-835 (1983)

1343  **Greene,R.L., Bajaj,K.K.**
TR    Energy levels of hydrogenic impurities and Wannier
      excitons in quantum well structures
      J. Vac. Sci. Technol. B 1   391-397 (1983)

1344  **Grubin,H.L., Kreskovsky,J.P.**
TR    The role of boundaries on high speed compound
      semiconductor devices
      Surf. Sci. 132   594-622 (1983)

1345  **Grunthaner,F.J., Madhukar,A.**
R     Growth, characterization, and properties of
      metastable and modulated semiconductor structures:
      prospects for future studies
      J. Vac. Sci. Technol. B 1   462-467 (1983)

1346  **Guldner,Y.**
R     Electronic properties of InAs-GaSb superlattices
      Physica 117B & 118B   735-740 (1983)

1347  **Hajdu,J.**
TR    Some remarks on the present understanding of the
      quantized Hall effect in two dimensions
      Lect. Notes Phys. 177   23-32 (1983)

1348  **Haldane,F.D.M.**
T     Fractional quantization of the Hall effect: a
      hierarchy of incompressible quantum fluid states
      Phys. Rev. Lett. 51   605-608 (1983)

1349  **Halperin,B.I.**
TR    Theory of the quantized Hall conductance
      Helv. Phys. Acta 56   75-102 (1983)

1350  **Harris,J.J., Woodcock,J.M.**
      The growth of thin, heavily doped layers for hot
      electron devices
      J. Vac. Sci. Technol. B 1   196-198 (1983)

1351  **Harris,J.S.,Jr., Asbeck,P.M., Miller,D.L.**
R     Heterojunction bipolar transistors
      Jpn. J. Appl. Phys. 22, Suppl. 22-1   375-380 (1983)

1352  **Hayes,J.R., Capasso,F., Gossard,A.C., Malik,R.J.,
      Wiegmann,W.**
      Bipolar transistor with graded band-gap base
      Electron. Lett. 19   410-411 (1983)

1353  **Hayes,J.R., Capasso,F., Malik,R.J., Gossard,A.C.,
      Wiegmann,W.**
      Optimum emitter grading for heterojunction bipolar
      transistors
      Appl. Phys. Lett. 43   949-951 (1983)

1354   Hayes,J.R., Capasso,F., Malik,R.J., Gossard,A.C.,
         Wiegmann,W.
       Elimination of the emitter/collector offset voltage
         in heterojunction bipolar transistors
       IEDM 83 Technical Digest    686-688 (1983)

1355   Heckingbottom,R., Davies,G.J., Prior,K.A.
  R    Growth and doping of gallium arsenide using molecular
         beam epitaxy (MBE): thermodynamic and kinetic
         aspects
       Surf. Sci. 132    375-389 (1983)

1356   Heiblum,M., Wang,W.I., Osterling,L.E., Deline,V.
       Heavy doping of GaAs and AlGaAs with silicon by
         molecular beam epitaxy
       J. Appl. Phys. 54    6751-6753 (1983)

1357   Heinonen,O., Taylor,P.L.
  T    Conductance plateaus in the quantized Hall effect
       Phys. Rev. B 28    6119-6122 (1983)

1358   Herman,M.A.
  R    Morphological stability in epitaxy of monolithically
         integrated optical devices
       Opt. Applicata 13    55-67 (1983)

1359   Hernandez-Calderon,I., Hoechst,H.
       New method for the analysis of reflection high-energy
         electron diffraction: alpha-Sn(001) and InSb(001)
         surfaces
       Phys. Rev. B 27    4961-4964 (1983)

1360   Hess,K.
  R    Electron transport in heterojunctions and
         superlattices
       Physica 117B & 118B    723-728 (1983)

1361   Hesto,P.
  TR   Injection dependence of quasiballistic transport in
         GaAs at 77 K
       Surf. Sci. 132    623-636 (1983)

1362   Hoechst,H., Hernandez-Calderon,I.
       Angular resolved photoemission of InSb(001) and
         heteroepitaxial films of alpha-Sn(001)
       Surf. Sci. 126    25-31 (1983)

1363   Hoepfel,R.A., Gornik,E., Gossard,A.C., Wiegmann,W.
       Plasmon excitation by Coulomb scattering of electrons
         in 2D systems
       Physica 117B & 118B    646-648 (1983)

1364   Holah,G.D., Meeks,E.L., Eisele,F.L.
       Molecular beam epitaxial growth of InGaAsP
       J. Vac. Sci. Technol. B 1    182-185 (1983)

1365 Hong,C.S., Wang,W.I., Eastman,L.F., Wood,C.E.C.
Interface states and current threshold of
   GaAs/Al(x)Ga(1-x)As double-heterostructure lasers
   grown by molecular beam epitaxy
Inst. Phys. Conf. Ser. 65    297-302 (1983)

1366 Houghton,A.J.N., Andrews,D.A., Davies,G.J.,
   Ritchie,S.
Low-loss optical waveguides in MBE-grown GaAs/GaAlAs
   heterostructures
Opt. Commun. 46    164-166 (1983)

1367 Hove,J.M. van, Cohen,P.I., Lent,C.S.
Disorder on GaAs(001) surfaces prepared by molecular
   beam epitaxy
J. Vac. Sci. Technol. A 1    546-550 (1983)

1368 Hove,J.M. van, Lent,C.S., Pukite,P.R., Cohen,P.I.
Damped oscillations in reflection high energy
   electron diffraction during GaAs MBE
J. Vac. Sci. Technol. B 1    741-746 (1983)

1369 Hove,J.M. van, Pukite,P.R., Cohen,P.I., Lent,C.S.
RHEED streaks and instrument response
J. Vac. Sci. Technol. A 1    609-613 (1983)

1370 Hsieh,K.H., Hollis,M., Wicks,G.W., Wood,C.E.C.,
   Eastman,L.F.
Ohmic contact behavior of Al metal epitaxy on GaInAs
   by MBE
Inst. Phys. Conf. Ser. 65    165-172 (1983)

1371 Hsieh,K.H., Ohno,H., Wicks,G.W., Eastman,L.F.
Dependence of electron mobility on spacer thickness
   and electron density in modulation-doped
   Ga(0.47)In(0.53)As/Al(0.48)In(0.52)As
   heterojunctions
Electron. Lett. 19    160-162 (1983)

1372 Hwang,J.C.M., Brennan,T.M., Cho,A.Y.
Initial results of a high throughput MBE sytem for
   device fabrication
J. Electrochem. Soc. 130    493-496 (1983)

1373 Hwang,J.C.M., Temkin,H., Brennan,T.M., Frahm,R.E.
Growth of high-purity GaAs layers by molecular beam
   epitaxy
Appl. Phys. Lett. 42    66-68 (1983)

1374 Iafrate,G.J., Malik,R.J., Tang,J.Y., Hess,K.
   T  Transient transport and transferred electron
         behaviour in gallium arsenide under the condition
         of high-energy electron injection
      Solid State Commun. 45    255-258 (1983)

1375   Inoue,M., Hida,H., Inayama,M., Inuishi,Y.,
       Nanbu,K., Hiyamizu,S.
       2D hot electron transport in a modulation-doped
          GaAs/AlGaAs interface
       Physica 117B & 118B   720-722 (1983)

1376   Inoue,M., Hiyamizu,S., Inayama,M., Inuishi,Y.
  R    Analysis of 2D electron transport at a GaAs/AlGaAs
          interface
       Jpn. J. Appl. Phys. 22, Suppl. 22-1   357-363 (1983)

1377   Inoue,M., Inayama,M., Hiyamizu,S., Inuishi,Y.
       Parallel electron transport and field effects of
          electron distributions in selectively-doped
          GaAs/n-AlGaAs
       Jpn. J. Appl. Phys. 22   L213-L215 (1983)

1378   Isihara,A.
  T    Theory of the quantized Hall conductivity plateaus of
          two-dimensional electron systems
       Solid State Commun. 46   265-267 (1983)

1379   Iwamura,H., Saku,T., Ishibashi,T., Otsuka,K.,
       Horikoshi,Y.
       Dynamic behaviour of a GaAs-AlGaAs MQW laser diode
       Electron. Lett. 19   180-181 (1983)

1380   Iwamura,H., Saku,T., Kobayashi,H., Horikoshi,Y.
       Spektrum studies on a GaAs-AlGaAs multi-quantum-well
          laser diode grown by molecular beam epitaxy
       J. Appl. Phys. 54   2692-2695 (1983)

1381   Jacobi,K.
  R    Electronic properties and surface geometry of GaAs
          and ZnO surfaces
       Surf. Sci. 132   1-21 (1983)

1382   Jewell,J.L., Gibbs,H.M., Gossard,A.C., Passner,A.,
       Wiegmann,W.
       Fabrication of GaAs bistable optical devices
       Mater. Lett. 1   148-151 (1983)

1383   Jung,H., Doehler,G.H., Goebel,E.O., Ploog,K.
       Optical gain in GaAs doping superlattices
       Appl. Phys. Lett. 43   40-42 (1983)

1384   Jung,H., Kuenzel,H., Doehler,G.H., Ploog,K.
       Photoluminescence in GaAs doping superlattices
       J. Appl. Phys. 54   7211-7220 (1983)

1385   Kastalsky,A., Dingle,R., Cheng,K.Y., Cho,A.Y.
       Two-dimensional electron gas at an MBE-grown,
          selectively-doped,
             In(0.53)Ga(0.47)As/In(0.52)Al(0.48)As interface
       Inst. Phys. Conf. Ser. 65   181-185 (1983)

**1386 Kastalsky,A., Luryi,S.**
T Novel real-space hot-electron transfer devices
IEEE Electron Device Lett. EDL-4   334-336 (1983)

**1387 Kaushik,S.B., Purohit,R.K., Sharma,B.L.**
T Detectivity calculations of III-V ternary compound PDB photodetectors
Infrared Phys. 23   15-18 (1983)

**1388 Kawabe,M., Kondo,M., Matsuura,N., Yamamoto,K.**
Photoluminescence of $Al(x)Ga(1-x)As/Al(y)Ga(1-y)As$ multiquantum wells grown by pulsed molecular beam epitaxy
Jpn. J. Appl. Phys. 22   L64-L66 (1983)

**1389 Kawamura,Y., Asahi,H.**
Silicon doping in InP grown by molecular beam epitaxy
Appl. Phys. Lett. 43   780-782 (1983)

**1390 Kawamura,Y., Asahi,H., Nagai,H.**
Electrical and optical properties of Be-doped InP grown by molecular beam epitaxy
J. Appl. Phys. 54   841-846 (1983)

**1391 Kawamura,Y., Asahi,H., Nagai,H., Ikegami,T.**
0.66 um room-temperature operation of InGaAlP DH laser diodes grown by MBE
Electron. Lett. 19   163-165 (1983)

**1392 Kelly,M.J.**
T Carrier statistics and the low-temperature mobility of semiconductor superlattices
J. Phys. C 16   L1165-L1168 (1983)

**1393 Khapachev,Y.P.**
T The theory of dynamical X-ray diffraction on a superlattice
Phys. Status Solidi B 120   155-163 (1983)

**1394 Khapachev,Y.P., Kuznetsov,G.F.**
T Dynamic diffraction of x rays in a harmonic superlattice
Sov. Phys. Crystallogr. 28   12-14   *
   Kristallografiya 28   27-31 (1983)

**1395 Kiehl,R.A., Feuer,M.D., Hendel,R.H., Hwang,J.C.M., Keramidas,V.G., Allyn,C.L., Dingle,R.**
Selectively doped heterostructure frequency dividers
IEEE Electron Device Lett. EDL-4   377-379 (1983)

**1396 Kleinman,D.A.**
T Binding energy of biexcitons and bound excitons in quantum wells
Phys. Rev. B 28   871-879 (1983)

1397 Klem,J., Fischer,R., Drummond,T.J., Morkoc,H., Cho,A.Y.
Incorporation of Sb in GaAs(1-x)Sb(x) (x<0.15) by molecular beam epitaxy
Electron. Lett. 19 453-455 (1983)

1398 Klem,J., Masselink,W.T., Arnold,D., Fischer,R., Drummond,T.J., Morkoc,H., Lee,K., Shur,M.S.
Persistent photoconductivity in (Al,Ga)As/GaAs modulation doped structures: dependence on structure and growth temperature
J. Appl. Phys. 54 5214-5217 (1983)

1399 Klitzing,K. von
R Quantized Hall effect
J. Magn. Magn. Mater. 31-34 525-529 (1983)

1400 Klitzing,K. von, Ebert,G.
R The quantum Hall effect
Physica 117B & 118B 682-687 (1983)

1401 Klitzing,K. von, Tausendfreund,B., Obloh,H., Herzog,T.
R Precision determination of $h/e^{**2}$ and the fine-structure constant from magneto-transport measurements on 2D electronic systems
Lect. Notes Phys. 177 1-10 (1983)

1402 Kobayashi,H., Iwamura,H., Saku,T., Otsuka,K.
Polarisation-dependent gain-current relationship in GaAs-AlGaAs MQW laser diodes
Electron. Lett. 19 166-168 (1983)

1403 Kondo,K., Muto,S., Nanbu,K., Ishikawa,T., Hiyamizu,S., Hashimoto,H.
Effect of H2 on the quality of Si-doped Al(x)Ga(1-x)As grown by MBE
Jpn. J. Appl. Phys. 22 L121-L123 (1983)

1404 Kroemer,H.
R Heterostructure bipolar transistors: What should we build?
J. Vac. Sci. Technol. B 1 126-130 (1983)

1405 Kroemer,H.
R Heterostructure devices: a device physicist looks at interfaces
Surf. Sci. 132 543-576 (1983)

1406 Kroemer,H., Griffiths,G.
T Staggered-lineup heterojunctions as sources of tunable below-gap radiation: operating principle and semiconductor selection
IEEE Electron Device Lett. EDL-4 20-22 (1983)

**1407** Krusor,B.S., Bachrach,R.Z.
Two-stage arsenic cracking source with integral getter pump for MBE growth
J. Vac. Sci. Technol. B 1   138-141 (1983)

**1408** Kuan,T.S., Chang,C.A.
Electron microscope studies of a Ge-GaAs superlattice grown by molecular beam epitaxy
J. Appl. Phys. 54   4408-4413 (1983)

**1409** Kuenzel,H., Fischer,A., Knecht,J., Ploog,K.
A new semiconductor superlattice with tunable electronic properties and simultaneously with mobility enhancement of electrons and holes
Appl. Phys. A 30   73-81 (1983)

**1410** Kuenzel,H., Fischer,A., Knecht,J., Ploog,K.
Investigation of persistent photoconductivity in Si-doped n-$Al(x)Ga(1-x)As$ grown by molecular beam epitaxy
Appl. Phys. A 32   69-78 (1983)

**1411** Kuramoto,Y.
T   Comment on "Two-dimensional magnetotransport in the extreme quantum limit"
Phys. Rev. Lett. 50   866 (1983)

**1412** Kuroda,T., Iwakuro,H., Nakamura,S.
Basic research on molecular beam epitaxy
Mem. Inst. Sci. Ind. Res., Osaka Univ., 40   25-45 (1983)

**1413** Lai,W.Y., Shen,J.L., Su,Z.B., Yu,L.
T   Quantized Hall effect in very strong magnetic field
Commun. Theor. Phys. 2   929-933 (1983)

**1414** Laidig,W.D., Caldwell,P.J., Kim,K., Lee,J.W.
All-binary AlAs-GaAs laser diode
IEEE Electron Device Lett. EDL-4   212-214 (1983)

**1415** Laidig,W.D., Lee,J.W., Wortman,J.J., Littlejohn,M.A.
AlAs-GaAs superlattices for optimum photoluminescence intensity
J. Vac. Sci. Technol. B 1   155-157 (1983)

**1416** Lambert,B., Deveaud,B., Regreny,A., Talalaeff,G.
Impurity photoluminescence in $GaAs/Ga(1-x)Al(x)As$ multiple quantum wells
Physica 117B & 118B   717-719 (1983)

**1417** Larsen,P.K., Neave,J.H., Veen,J.F. van der, Dobson,P.J., Joyce,B.A.
GaAs(001)-c(4x4): A chemisorbed structure
Phys. Rev. B 27   4966-4977 (1983)

1418 Larsen,P.K., Veen,J.F. van der
R    Photoemission from MBE grown III-V surfaces and
        interfaces
     Surf. Sci. 126    1-19 (1983)

1419 Lassnig,R., Gornik,E.
T    Calculations of the cyclotron resonance linewidth in
        GaAs-AlGaAs heterostructures
     Solid State Commun. 47    959-964 (1983)

1420 Lassnig,R., Zawadzki,W.
T    Theory of the magnetophonon effect for the
        two-dimensional electron gas in semiconducting
        heterostructures
     J. Phys. C 16    5435-5440 (1983)

1421 Laughlin,R.B.
T    Anomalous quantum Hall effect: an incompressible
        quantum fuid with fractionally charged excitations
     Phys. Rev. Lett. 50    1395-1398 (1983)

1422 Leburton,J.P., Hess,K.
T    A simple model for the index of refraction of
        GaAs-AlAs superlattices and  hetrostructure
        layers: contributions of the states around GAMMA
     J. Vac. Sci. Technol. B 1    415-419 (1983)

1423 Lee,C.P., Hou,D., Lee,S.J., Miller,D.L.,
        Anderson,R.J.
     Ultra high speed digital integrated circuits using
        GaAs/GaAlAs high electron mobility transistors
     1983 GaAs IC Symposium Technical Digest    162-165
        (1983)

1424 Lee,C.P., Wang,W.I.
     High-performance modulation-doped GaAs integrated
        circuits with planar structures
     Electron. Lett. 19    155-157 (1983)

1425 Lee,J.
T    Direct intersubband optical absorption of
        semiconducting thin wire
     J. Appl. Phys. 54    5482-5484 (1983)

1426 Lee,J., Spector,H.N.
T    Impurity-limited mobility of semiconductor thin wire
     J. Appl. Phys. 54    3921-3925 (1983)

1427 Lee,J., Spector,H.N., Arora,V.K.
T    Quantum transport in a single layered structure for
        impurity scattering
     Appl. Phys. Lett. 42    363-365 (1983)

1428 Lee,K., Shur,M.S., Drummond,T.J., Morkoc,H.
T   Current-voltage and capacitance-voltage characteristics of modulation-doped field-effect transistors
    IEEE Trans. Electron Devices ED-30   207-212 (1983)

1429 Lee,K., Shur,M.S., Drummond,T.J., Morkoc,H.
T   Electron density of the two-dimensional electron gas in modulation doped layers
    J. Appl. Phys. 54   2093-2096 (1983)

1430 Lee,K., Shur,M.S., Drummond,T.J., Morkoc,H.
    Low field mobility of 2-d electron gas in modulation doped $Al(x)Ga(1-x)As/GaAs$ layers
    J. Appl. Phys. 54   6432-6438 (1983)

1431 Lee,K., Shur,M.S., Drummond,T.J., Su,S.L., Lyons,W.G., Fischer,R., Morkoc,H.
    Design and fabrication of high transconductance modulation-doped (Al,Ga)As/GaAs FETs
    J. Vac. Sci. Technol. B 1   186-189 (1983)

1432 Lee,S.J., Crowell,C.R., Lee,C.P.
    Optimization of HEMTs in ultra high speed GaAs integrated circuits
    IEDM 83 Technical Digest   103-106 (1983)

1433 Levine,B.F., Bethea,C.G., Tsang,W.T., Capasso,F., Thornber,K.K., Fulton,R.C., Kleinman,D.A.
    Measurement of high electron drift velocity in a submicron, heavily doped graded gap $Al(x)Ga(1-x)As$ layer
    Appl. Phys. Lett. 42   769-771 (1983)

1434 Levine,B.F., Logan,R.A., Tsang,W.T., Bethea,C.G., Merritt,F.R.
    Optically integrated coherently coupled $Al(x)Ga(1-x)As$ lasers
    Appl. Phys. Lett. 42   339-341 (1983)

1435 Levine,H., Libby,S.B., Pruisken,A.M.M.
T   Electron delocalization by a magnetic field in two dimensions
    Phys. Rev. Lett. 51   1915-1918 (1983)

1436 Li,A.Z., Cheng,H., Milnes,A.G.
    Aspects of GaAs selective area growth by molecular beam epitaxy with patterning by $SiO_2$ masking
    J. Electrochem. Soc. 130   2072-2075 (1983)

1437 Li,A.Z., Milnes,A.G.
    The residual effects of germanium as an n-type dopant for GaAs during molecular beam epitaxial growth
    J. Cryst. Growth 62   95-105 (1983)

1438    Li,A.Z., Milnes,A.G.
         The residual effects of germanium as an n-type dopant
            for GaAs during molecular beam epitaxial growth
         J. Cryst. Growth 62    95-105 (1983)

1439    Li,A.Z., Xin,S.H., Milnes,A.G.
         Electrical and photoluminescence properties of
            Ge-doped n-type GaAs grown by molecular beam
            epitaxy
         J. Electron. Mater. 12    71-91 (1983)

1440    Lindemann,G., Seidenbusch,W., Lassnig,R.,
            Edlinger,J., Gornik,E.
         Cyclotron resonance studies of polarons and screening
            effects in GaAs
         Physica 117B & 118B    649-651 (1983)

1441    Linh,N.T.
     R   Applications of superlattices
         Helv. Phys. Acta 56    361-370 (1983)

1442    Linh,N.T.
     R   The two-dimensional electron gas and its technical
            application
         in "Festkoerperprobleme", Ed. P. Grosse (Vieweg,
            Braunschweig, 1983) Vol. XXIII    227-246 (1983)

1443    Littlejohn,M.A., Kwapien,W.M., Glisson,T.H.,
            Hauser,J.R., Hess,K.
     T   Effects of band bending on real space transfer in
            GaAs-Al(x)Ga(1-x)As layered heterostructures
         J. Vac. Sci. Technol. B 1    445-448 (1983)

1444    Littlejohn,M.A., Trew,R.J., Hauser,J.R.,
            Golio,J.M.
     T   Electron transport in planar-doped barrier structures
            using an ensemble Monte Carlo method
         J. Vac. Sci. Technol. B 1    449-454 (1983)

1445    Livingstone,A., Tsubaki,K., Kawashima,M.,
            Okamoto,H., Kumabe,K.
         Field ionised impurity scattering in an AlGaAs/GaAs
            two-dimensional electron gas
         Electron. Lett. 19    619-620 (1983)

1446    Loreck,L., Daembkes,H., Heime,K., Ploog,K.
         Deep level analysis in (AlGa)As-GaAs MODFETs by means
            of low frequency noise measurements
         IEDM 83 Technical Digest    107-110 (1983)

1447    Low,T.S., Skromme,B.J., Stillman,G.E.
         Incorporation of amphoteric impurities in high purity
            GaAs
         Inst. Phys. Conf. Ser. 65    515-522 (1983)

**1448  Ludeke,R.**
R   The formation of interfaces on GaAs and related
       semiconductors: a reassessment
    Surf. Sci. 132   143-168 (1983)

**1449  Ludeke,R., Chiang,T.C., Eastman,D.E.**
    Core-level photoemission studies of MBE-grown
       semiconductor surfaces
    Physica 117B & 118B   819-821 (1983)

**1450  Luryi,S., Kazarinov,R.F.**
T   Theory of quantized Hall effect at low temperatures
    Phys. Rev. B 27   1386-1389 (1983)

**1451  Lyons,W.G., Arnold,D., Thorne,R.E., Su,S.L.,
       Kopp,W., Morkoc,H.**
    Normally-on and normally-off camel diode gate GaAs
       and modulation doped GaAs/Al(x)Ga(1-x)As field
       effect transistors
    Inst. Phys. Conf. Ser 65   379-384 (1983)

**1452  Maan,J.C.**
R   Magneto-optical experiments on a thin InAs layer
       confined between GaSb in a parallel and
       perpendicular magnetic field
    Lect. Notes Phys. 177   163-173 (1983)

**1453  Maan,J.C., Englert,T., Uihlein,C., Kuenzel,H.,
       Fischer,A., Ploog,K.**
    Quantum transport of electrons in a thin GaAs layer
       by an impurity space charge potential in high
       magnetic fields
    Solid State Commun. 47   383-386 (1983)

**1454  MacDonald,A.H.**
T   Quantized Hall conductance in a relativistic
       two-dimensional electron gas
    Phys. Rev. B 28   2235-2236 (1983)

**1455  MacDonald,A.H., Rice,T.M., Brinkman,W.F.**
T   Hall voltage and current distributions in an ideal
       two-dimensional system
    Phys. Rev. B 28   3648-3650 (1983)

**1456  MacKinnon,A.**
T   The effect of disorder on the quantised Hall
       conductivity
    J. Phys. C 16   L945-L948 (1983)

**1457  Madhukar,A.**
TR  Far from equilibrium vapour phase growth of lattice
       matched III-V compound semiconductor interfaces:
       some basic concepts and Monte-Carlo computer
       simulations
    Surf. Sci. 132   344-374 (1983)

1458  Mailhiot,C., McGill,T.C., Schulman,J.N.
T     Tunneling and propagation transport in
      GaAs-Ga(1-x)Al(x)As-GaAs(100) double
      heterojunctions
      J. Vac. Sci. Technol. B 1   439-444 (1983)

1459  Mailhiot,C., Smith,D.L., McGill,T.C.
T     Transport characteristics of L-point and Gamma-point
      electrons through GaAs-Ga(1-x)Al(x)As-GaAs(111)
      double heterojunctions
      J. Vac. Sci. technol. B 1   637-642 (1983)

1460  Maki,K., Zotos,X.
T     Static and dynamic properties of a two-dimensional
      Wigner crystal in a strong magnetic field
      Phys. Rev. B 28   4349-4356 (1983)

1461  Maki,P.A., Palmateer,S.C., Wicks,G.W.,
        Eastman,L.F., Calawa,A.R.
      Effect of substrate annealing and V:III flux ratio on
      the molecular beam epitaxial growth of AlGaAs-GaAs
      single quantum wells
      J. Electron. Mater. 12   1051-1063 (1983)

1462  Malik,R.J., Hayes,J.R., Capasso,F., Alavi,K.,
        Cho,A.Y.
      High-gain Al(0.48)In(0.52)As/Ga(0.47)In(0.53)As
      vertical n-p-n heterojunction bipolar transistors
      grown by molecular beam epitaxy
      IEEE Electron Device Lett. EDL-4   383-385 (1983)

1463  Markiewicz,R.S., Widom,A., Sokoloff,J.
T     Anomalous quantum Hall effect - origin of fractional
      Hall steps
      Phys. Rev. B 28   3654-3655 (1983)

1464  Marzin,J.Y., Rao,E.V.K.
      Optical studies of In(x)Ga(1-x)As-GaAs strained
      multiquantum well structures
      Appl. Phys. Lett. 43   560-562 (1983)

1465  Massies,J., Sauvage-Simkin,M.
      Mismatch and electron mobility in MBE Ga(x)In(1-x)As
      epitaxial layers on InP substrates
      Appl. Phys. A 32   27-30 (1983)

1466  Matsubara,K., Takagi,T.
      Film growth of GaN on a c-axis oriented ZnO film
      using reactive ionized-cluster beam technique and
      its application to thin film devices
      Jpn. J. Appl. Phys. 22, Suppl. 22-1   511-514 (1983)

1467  Mendez,E.E., Bastard,G., Chang,L.L., Esaki,L.
      Electric field-induced quenching of luminescence in
      quantum wells
      Physica 117B & 118B   711-713 (1983)

1468 Mendez,E.E., Chang,C.A., Takaoka,H., Chang,L.L., Esaki,L.
Optical properties of GaSb-AlSb superlattices
J. Vac. Sci. Technol. B 1   152-154 (1983)

1469 Mendez,E.E., Heiblum,M., Chang,L.L., Esaki,L.
High-magnetic-field transport in a dilute two-dimensional electron gas
Phys. Rev. B 28   4886-4888 (1983)

1470 Mendez,E.E., Heiblum,M., Fischer,R., Klem,J., Thorne,R.E., Morkoc,H.
Photoluminescence study of the incorporation of silicon in GaAs grown by molecular beam epitaxy
J. Appl. Phys. 54   4202-4204 (1983)

1471 Metze,G.M., Calawa,A.R.
Effects of very low growth rates on GaAs grown by molecular beam epitaxy at low substrate temperatures
Appl. Phys. Lett. 42   818-820 (1983)

1472 Metze,G.M., Calawa,A.R., Mavroides,J.G.
An investigation of GaAs films grown by MBE at low substrate temperatures and growth rates
J. Vac. Sci. Technol. B 1   166-169 (1983)

1473 Milano,R.A., Higgins,J.A., Miller,D.L., Sovero,E.A.
The application of modulation-doped heterostructures to high speed charge coupled devices
Inst. Phys. Conf. Ser. 65   445-452 (1983)

1474 Milanovic,V.
T   Determination of the discrete "surface" energy level in semi-infinite superlattice structures
Physica 121 B   181-186 (1983)

1475 Milanovic,V., Tjapkin,D.
T   Self-consistent evaluation of nonuniform superlattice parameters by the harmonic method
Physica 121 B   187-192 (1983)

1476 Miller,D.A.B.
R   Dynamic nonlinear optics in semiconductors: physics and applications
Laser Focus 19 (July 1983)   61-68 (1983)

1477 Miller,D.A.B., Chemla,D.S., Eilenberger,D.J., Smith,P.W., Gossard,A.C.
Degenerate four-wave mixing in room-temperature GaAs/GaAlAs multiple quantum well structures
Appl. Phys. Lett. 42   925-927 (1983)

1478 Miller,D.A.B., Chemla,D.S., Smith,P.W.,
Gossard,A.C., Wiegmann,W.
Nonlinear optics with a diode-laser light source
Opt. Lett. 8    477-479 (1983)

1479 Miller,D.L., Asbeck,P.M., Anderson,R.J.,
Eisen,P.H.
(GaAl)As/GaAs heterojunction bipolar transistors with
graded composition in the base
Electron. Lett. 19    367-368 (1983)

1480 Miller,R.C., Gossard,A.C.
Anomalous polarization in the photoluminescence from
Si-doped GaAs-Al(x)Ga(1-x)As quantum well samples
Phys. Rev. B 28    3645-3647 (1983)

1481 Miller,R.C., Gossard,A.C.
Some effects of a longitudinal electric field on the
photoluminescence of p-doped GaAs-Al(x)Ga(1-x)As
quantum well heterostructures
Appl. Phys. Lett. 43    954-956 (1983)

1482 Miller,R.C., Gossard,A.C., Tsang,W.T.
Extrinsic photoluminescence from GaAs quantum wells
Physica 117B & 118B    714-716 (1983)

1483 Mimura,T.
R  High electron mobility transistor for VLSI
Jpn. Annu. Rev. Electron., Comput. & Telecommun.:
Semiconductor Technology 8    277-294 (1983)

1484 Mimura,T.
R  Why HEMT are necessary and how they are made
JEE, J. Electron. Eng. 20    60-62 (1983)

1485 Mimura,T., Joshin,K., Kuroda,S.
Device modeling of HEMTs
Fujitsu Sci. Tech. J. 19    243-278 (1983)

1486 Mimura,T., Nishiuchi,K., Abe,M., Shibatomi,A.,
Kobayashi,M.
High electron mobility transistors for LSI circuits
IEDM 83 Technical Digest    99-102 (1983)

1487 Nakagawa,T., Kawai,N.J., Ohta,K., Kawashima,M.
T  New negative-resistance device by a CHIRP superlattice
Electron. Lett. 19    822-823 (1983)

1488 Narita,S., Muro,K., Mori,S., Hiyamizu,S., Nanbu,K.
Anomalous cyclotron resonance of 2D electrons in
Al(x)Ga(1-x)As/GaAs heterojunctions
Lect. Notes Phys. 177    194-198 (1983)

1489  Nathan,M.I., Jackson,T.N., Kirchner,P.D.,
      Mendez,E.E., Pettit,G.D., Woodall,J.M.
      Persistent photo-conductance and photoquenching of
         selectively doped Al(0.3)Ga(0.7)/GaAs
         heterojunctions
      J. Electron. Mater. 12   719-725 (1983)

1490  Neave,J.H., Dobson,P.J., Harris,J.J., Dawson,P.,
      Joyce,B.A.
      Silicon doping of MBE-grown GaAs films
      Appl. Phys. A 32   195-200 (1983)

1491  Neave,J.H., Joyce,B.A., Dobson,P.J., Norton,N.
      Dynamics of film growth of GaAs by MBE from RHEED
         observations
      Appl. Phys. A 31   1-8 (1983)

1492  Neave,J.H., Larsen,P.K., Joyce,B.A., Gowers,J.P.,
      Veen,J.F. van der
      Some observations on Ge:GaAs(001) and GaAs:Ge(001)
         interfaces and films
      J. Vac. Sci. Technol. B 1   668-674 (1983)

1493  Neave,J.H., Larsen,P.K., Veen,J.F. van der,
      Dobson,P.J., Joyce,B.A.
      Effect of arsenic species (As2 or As4) on the
         crystallographic and electronic structure of
         MBE-grown GaAs(001) reconstructed surfaces
      Surf. Sci. 113   267-278 (1983)

1494  Neumann,D.A., Zabel,H., Morkoc,H.
      X-ray evidence for a terraced GaAs/AlAs superlattice
      Appl. Phys. Lett. 43   59-61 (1983)

1495  Nicholas,R.J., Brummell,M.A., Portal,J.C.,
      Cheng,K.Y., Cho,A.Y., Pearsall,T.P.
      An experimental determination of enhanced electron
         g-factors in GaInAs-AlInAs heterojunctions
      Solid State Commun. 45   911-914 (1983)

1496  Nicholas,R.J., Brummell,M.A., Portal,J.C.,
      Cho,A.Y., Cheng,K.Y., Pearsall,T.P.
      Quantum transport in GaInAs-AlInAs heterojunctions
      Lect. Notes Phys. 177   110-113 (1983)

1497  Niina,T., Yoneda,K., Toda,T.
      Dependence of the characteristics of ZnSe MBE grown
         on GaAs and GaP on thermal treatment in a vacuum
      J. Electrochem. Soc. 130   2099-2104 (1983)

1498  Nishi,H., Inada,T., Saito,J., Ishikawa,T.,
      Hiyamizu,S.
      Implantation into an AlGaAs/GaAs heterostructure
      Jpn. J. Appl. Phys. 22, Suppl. 22-1   401-404 (1983)

1499 Noreika,A.J., Greggi,J.,Jr., Takei,W.J.,
     Francombe,M.H.
     Properties of MBE grown InSb and $InSb_{(1-x)}Bi_{(x)}$
     J. Vac. Sci. Technol. A 1   558-561 (1983)

1500 Nottenburg,R., Buehlmann,H.J., Bischoff,J.C.,
     Ilegems,M.
     MBE $Al_{(x)}Ga_{(1-x)}As$/GaAs phototransistors sensitive at
        low illumination
     IEDM 83 Technical Digest   472-474 (1983)

1501 Oelhafen,P., Freeouf,J.L., Pettit,G.D.,
     Woodall,J.M.
     Elevated temperature low energy ion cleaning of GaAs
     J. Vac. Sci. Technol. B 1   787-790 (1983)

1502 Ogura,M., Hata,T., Kawai,N.J., Yao,T.
     GaAs/$Al_{(x)}Ga_{(1-x)}As$ multilayer reflector for surface
        emitting laser diode
     Jpn. J. Appl. Phys. 22   L112-L114 (1983)

1503 Okamoto,H., Seki,M., Horikoshi,Y.
     Direct observation of lattice arrangement in MBE
        grown GaAs-AlGaAs superlattices
     Jpn. J. Appl. Phys. 22   L367-L369 (1983)

1504 Olego,D., Chang,T.Y., Silberg,E., Caridi,E.A.,
     Pinczuk,A.
     Compositional dependence of band-gap energy,
        conduction-band effective mass and lattice
        vibrations of $In_{(1-x-y)}Ga_{(x)}Al_{(y)}As$ lattice
        matched to Inp
     Inst. Phys. Conf. Ser. 65   195-202 (1983)

1505 Olego,D., Pinczuk,A., Gossard,A.C., Wiegmann,W.
     Plasma oscillations of layered electron gases in
        semiconductor heterostructures
     J. Vac. Sci. Technol. B 1   412-414 (1983)

1506 Ono,Y.
   T Self-consistent treatment of dynamical diffusion
        coefficient of two dimensional random electron
        system under strong magnetic fields II
     J. Phys. Soc. Jpn. 52   2492-2498 (1983)

1507 Osbourn,G.C.
   T $In_{(x)}Ga_{(1-x)}As$-$In_{(y)}Ga_{(1-y)}As$ strained-layer
        superlattices: a proposal for useful, new
        electronic materials
     Phys. Rev. B 27   5126-5128 (1983)

1508 Osbourn,G.C.
  TR Electronic properties of strained-layer superlattices
     J. Vac. Sci. Technol. B 1   379-382 (1983)

1509 Paalanen,M.A., Tsui,D.C., Hwang,J.C.M.
Parabolic magnetoresistance from the interaction
effect in a two-dimensional electron gas
Phys. Rev. Lett. 51   2226-2229 (1983)

1510 Palmateer,S.C., Maki,P.A., Hollis,M., Eastman,L.F.,
Hitzman,C.J., Ward,I.
A study of substrate effects on planar doped
structures in gallium arsenide grown by molecular
beam epitaxy
Inst. Phys. Conf. Ser. 65   149-156 (1983)

1511 Palmateer,S.C., Schaff,W.J., Galuska,A.,
Berry,J.D., Eastman,L.F.
Heat treatment of semi-insulating chromium-doped
gallium arsenide substrates with converted surface
removed prior to molecular beam epitaxial growth
Appl. Phys. Lett. 42   183-185 (1983)

1512 Pang,Y.M., Chen,C.Y., Garbinski,P.A.
1.5 GHz operation of an Al(x)Ga(1-x)As/GaAs
modulation doped photoconductive detector
Electron. Lett. 19   716-717 (1983)

1513 Paparoditis,C., Rideau,A., Monnom,G., Gaucherel,P.
Disorder and optical properties in gallium arsenide
Physica 117B & 118B   992-994 (1983)

1514 Parayenthal,P., Ro,C.S., Pollak,F.H., Stanley,C.R.,
Wicks,G.W., Eastman,L.F.
Electroreflectance investigation of
(Ga(1-x)Al(x))(0.47)In(0.53)As lattice matched to
InP
Appl. Phys. Lett. 43   109-111 (1983)

1515 Pearsall,T.P., Hendel,R., O'Connor,P., Alavi,K.,
Cho,A.Y.
Selectively doped
Al(0.48)In(0.52)As/Ga(0.47)In(0.53)As
heterostructure field effect transistor
IEEE Electron Device Lett. EDL-4   5-8 (1983)

1516 People,R., Wecht,K.W., Alavi,K., Cho,A.Y.
Measurement of the conduction-band discontinuity of
molecular beam epitaxial grown
In(0.52)Al(0.48)As/In(0.53)Ga(0.47)As, N-n
heterojunction by C-V profiling
Appl. Phys. Lett. 43   118-120 (1983)

1517 Phelps,D.E., Bajaj,K.K.
T   Ground-state energy of a D- ion in two-dimensional
semiconductors
Phys. Rev. B 27   4883-4886 (1983)

1518    Phillips,J.M., Feldman,L.C., Gibson,J.M.,
            McDonald,M.L.
        Epitaxial growth of BaF2 on semiconductor substrates
        Thin Solid Films 104    101-107 (1983)

1519    Phillips,J.M., Feldman,L.C., Gibson,J.M.,
            McDonald,M.L.
        Epitaxial growth of alkaline earth fluorides on
            semiconductors
        Thin Solid Films 107    217-226 (1983)

1520    Phillips,J.M., Feldman,L.C., Gibson,J.M.,
            McDonald,M.L.
        Rutherford backscattering/channeling and transmission
            electron microscopy analysis of epitaxial BaF2
            films on Ge and InP
        J. Vac. Sci. Technol. B 1    246-249 (1983)

1521    Picraux,S.T., Dawson,L.R., Osbourn,G.C., Chu,W.K.
        Ion channeling studies of InGaAs/GaAs strained-layer
            superlattices
        Appl. Phys. Lett. 43    930-932 (1983)

1522    Pinczuk,A., Worlock,J.M.
  R     Light scattering by two-dimensional electron systems
        Physica 117B & 118B    637-642 (1983)

1523    Ploog,K., Doehler,G.H.
  R     Compositional and doping superlattices in III-V
            semiconductors
        Adv. Phys. 32    285-359 (1983)

1524    Ploog,K., Jung,H., Kuenzel,H., Ruden,P.
        Tunable photo- and electroluminescence from GaAs
            doping superlattices
        Jpn. J. Appl. Phys. 22, Suppl. 22-1    287-290 (1983)

1525    Pokatilov,E.P., Beril,S.I.
  T     Electron-phonon interaction in periodic two-layer
            structures
        Phys. Status Solidi B 118    567-573 (1983)

1526    Pollmann,J., Mazur,A.
  TR    Theory of semiconductor heterojunctions
        Thin Solid Films 104    257-276 (1983)

1527    Price,P.J.
  T     Hot electron effects in heterolayers
        Physica 117B & 118B    750-752 (1983)

1528    Price,P.J., Stern,F.
  TR    Carrier confinement effects
        Surf. Sci. 132    577-593 (1983)

1529 **Qin,G., Giuliani,G.F., Quinn,J.J.**
T   Acoustic surface plasmons in type-II semiconducting
        superlattices
    Phys. Rev. B 28    6144-6146 (1983)

1530 **Quinn,J.J., Strom,U., Chang,L.L.**
T   Direct electromagnetic generation of high frequency
        acoustic waves in semiconductor superlattices
    Solid State Commun. 45    111-112 (1983)

1531 **Rakshit,S., Chakraborti,N.B., Sarin,R.**
T   Multiplication noise in multi-heterostructure
        avalanche photodiodes
    Solid-State Electron. 26    999-1003 (1983)

1532 **Rammal,R., Toulouse,G., Jaekel,M.T., Halperin,B.I.**
T   Quantized Hall conductance and edge states:
        two-dimensional strips with a periodic potential
    Phys. Rev. B 27    5142-5145 (1983)

1533 **Ranke,W.**
    Ultraviolet photoelectron spectroscopy investigation
        of electron affinity and polarity on a cylindrical
        GaAs single crystal
    Phys. Rev. B 27    7808-7810 (1983)

1534 **Ranke,W.**
    Oxygen adsorption on a cylindrical GaAs single
        crystal prepared by molecular beam epitaxy
    Phys. Scr. T4    100-102 (1983)

1535 **Rehm,W., Kuenzel,H., Doehler,G.H., Ploog,K.,
        Ruden,P.**
    Time resolved luminescence in n-i-p-i doping
        superlattices
    Physica 117B & 118B    732-734 (1983)

1536 **Rehm,W., Ruden,P., Doehler,G.H., Ploog,K.**
    Study of time-resolved luminescence in GaAs doping
        superlattices
    Phys. Rev. B 28    5937-5942 (1983)

1537 **Reich,R.K., Grondin,R.O., Ferry,D.K.**
T   Transport in lateral surface superlattices
    Phys. Rev. B 27    3483-3493 (1983)

1538 **Ridley,B.K.**
T   On the possibility of intrinsic negative differential
        resistance in III-V quantum wells
    J. Phys. C 16    L789-L790 (1983)

1539 **Rochette,J.F., Delescluse,P., Laviron,M.,
        Delagebeaudeuf,D., Chevrier,J., Linh,N.T.**
    Low temperature persistent photoconductivity in
        two-dimensional electron gas FETs
    Inst. Phys. Conf. Ser. 65    385-392 (1983)

1540  Rowe,J.E., Pearsall,T.P., Logan,R.A.
      A study of the valence band structure of
          Ga(0.47)In(0.53)As by angle resolved photoemission
      Physica 117B & 118B   347-349 (1983)

1541  Ruden,P.
   T  Elementary excitations in semiconductors with n-i-p-i
          doping superlattices
      J. Vac. Sci. Technol. B 1   285-288 (1983)

1542  Ruden,P., Doehler,G.H.
   T  Semiconductors with hetero-n-i-p-i superlattices
      Surf. Sci. 132   540-542 (1983)

1543  Ruden,P., Doehler,G.H.
   T  Anisotropy effects and optical excitation of acoustic
          phonons in n-i-p-i doping superlattices
      Solid State Commun. 45   23-25 (1983)

1544  Ruden,P., Doehler,G.H.
   T  Electronic structure of semiconductors with doping
          superlattices
      Phys. Rev. B 27   3538-3546 (1983)

1545  Ruden,P., Doehler,G.H.
   T  Electronic excitations in semiconductors with doping
          superlattices
      Phys. Rev. B 27   3547-3553 (1983)

1546  Ryuzan,O., Misugi,T.
   R  New developments in III-V transistor technology
      Physica 117 B & 118 B   50-54 (1983)

1547  Saito,J., Nanbu,K., Ishikawa,T., Hiyamizu,S.
      Dependence of the mobility and the concentration of
          two-dimensional electron gas in selectively doped
          GaAs/n-Al(x)Ga(1-x)As heterostructures on the AlAs
          mole fraction
      Jpn. J. Appl. Phys. 22   L79-L81 (1983)

1548  Sakaki,H., Ohno,H., Nishi,S., Yoshino,J.
      Effect of tangential magnetic field on the
          two-dimensional electron transport in
          n-AlGaAs/GaAs superlattices and hetero-interfaces
      Physica 117B & 118B   703-705 (1983)

1549  Salmon,L.G., Rhodin,T.N.
      Angle-resolved photoemission study of GaAs(100)
          surfaces grown by molecular beam epitaxy
      J. Vac. Sci. Technol. B 1   736 (1983)

1550  Salmon,L.G., Rhodin,T.N.
      Angle-resolved photoemission study of the
          Ge/GaAs(100) interface
      Inst. Phys. Conf. Ser. 65   561-568 (1983)

1551  Sapriel,J., Djafari-Rouhani,B., Dobrzynski,L.
T     Vibrations in superlattices; application to GaAs-AlAs
         systems
      Surf. Sci. 126    197-201 (1983)

1552  Sapriel,J., Michel,J.C., Toledano,J.C., Vacher,R.,
         Kervarec,J., Regreny,A.
      Light scattering from vibrational modes in
         GaAs-Ga(1-x)Al(x)As superlattices and related
         alloys
      Phys. Rev. B 28    2007-2016 (1983)

1553  Saunier,P., Shih,H.D.
      High-performance K-band GaAs power field-effect
         transistors prepared by molecular beam epitaxy
      Appl. Phys. Lett. 42    966-968 (1983)

1554  Schaffer,W.J., Lind,M.D., Kowalczyk,S.P.,
         Grant,R.W.
      Nucleation and strain relaxation at the
         InAs/GaAs(100) heterojunction
      J. Vac. Sci. Technol. B 1    688-695 (1983)

1555  Schirber,J.E., Fritz,I.J., Dawson,L.R.,
         Osbourn,G.C.
      Quantum oscillations in strained-layer superlattices
      Phys. Rev. B 28    2229-2231 (1983)

1556  Schmidt,R.R., Bosman,G., Vliet,C.M. van,
         Eastman,L.F., Hollis,M.
      Noise in near-ballistic n+nn+ and n+pn+ gallium
         arsenide submicron diodes
      Solid-State Electron. 26    437-444 (1983)

1557  Schneider,M.V., Cho,A.Y., Kollberg,E., Zirath,H.
      Characteristic of Schottky diodes with microcluster
         interface
      Appl. Phys. Lett. 43    558-560 (1983)

1558  Schulman,J.N.
T     Effects of compositional grading on
         GaAs-Ga(1-x)Al(x)As interface and quantum well
         electronic structure
      J. Vac. Sci. Technol. B 1    644-647 (1983)

1559  Shah,J., Pinczuk,A., Stoermer,H.L., Gossard,A.C.,
         Wiegmann,W.
      Electric field induced heating of high mobility
         electrons in modulation doped GaAs - AlGaAs
         heterostructures
      Appl. Phys. Lett. 42    55-57 (1983)

1560   Shank,C.V., Fork,R.L., Yen,R., Shah,J.,
           Greene,B.I., Gossard,A.C., Weisbuch,C.
       Picosecond dynamics of hot carrierrelaxation in
           highly excited multi-quantum well structures
       Solid State Commun. 47   981-983 (1983)

1561   Shih,H.D., Bayraktaroglu,B., Duncan,W.M.
       Growth of millimeter-wave GaAs IMPATT structures by
           molecular beam epitaxy
       J. Vac. Sci. Technol. B 1   199-201 (1983)

1562   Shinozuka,Y., Matsuura,M.
   T   Wannier excitation in quantum wells
       Phys. Rev. B 28   4878-4881 (1983)

1563   Shmelev,G.M., Shon,N.H., Tsurkan,G.I.
   T   Giant quantum oscillations of sound absorption in a
           superlattice
       J. Phys. C 16   2587-2590 (1983)

1564   Shmelev,G.M., Son,N.H., Anh,V.H.
   T   Sound absorption in semiconductors with a
           superlattice in a quantizing magnetic field
       Solid State Commun. 48   239-242 (1983)

1565   Silberg,E., Chang,T.Y., Caridi,E.A.,
           Evans,C.A.,Jr., Hitzman,C.J.
       Manganese and germanium redistribution in
           In(0.53)Ga(0.47)As grown by molecular beam epitaxy
       J. Vac. Sci. Technol. B 1   178-181 (1983)

1566   Silberg,E., Chang,T.Y., Caridi,E.A.,
           Evans,C.A.,Jr., Hitzman,C.J.
       Spatially correlated redistribution of Mn and Ge in
           In(1-x)Ga(x)As MBE layers
       Inst. Phys. Conf. Ser. 65   187-194 (1983)

1567   Singer,P.H.
   R   Molecular beam epitaxy for research and production
       Semiconductor International No. 10 (October 1983)
           72-80 (1983)

1568   Singh,J., Madhukar,A.
   T   Surface orientation dependent surface kinetics and
           interface roughening in molecular beam epitaxial
           growth of III-V semiconductors: a Monte Carlo study
       J. Vac. Sci. Technol. B 1   305-312 (1983)

1569   Singh,J., Madhukar,A.
   T   Prediction of kinetically controlled surface
           roughening: a Monte Carlo computer-simulation study
       Phys. Rev. Lett. 51   794-797 (1983)

1570 Slater,N.J., Choudhury,A.N.M.M.,
 Tabatabaie-Alavi,K., Rowe,W., Fonstad,C.G.,
 Alavi,K., Cho,A.Y.
 Ion implantation doping of InGaAs and InAlAs
 Inst. Phys. Conf. Ser. 65   627-634 (1983)

1571 Smith,J.S., Chiu,L.C., Margalit,S., Yariv,A.,
 Cho,A.Y.
 A new infrared detector using electron emission from
  multilple quantum wells
 J. Vac. Sci. Technol. B 1   376-378 (1983)

1572 Sollner,T.C.L.G., Goodhue,W.D., Tannenwald,P.E.,
 Parker,C.D., Peck,D.D.
 Resonant tunneling through quantum wells at
  frequencies up to 2.5 THz
 Appl. Phys. Lett. 43   588-590 (1983)

1573 Solomon,P.M., Hickmott,T.W., Morkoc,H., Fischer,R.
 Electrical measurements on n+GaAs-undoped
  Ga(0.6)Al(0.4)As-n-GaAs capacitors
 Appl. Phys. Lett. 42   821-823 (1983)

1574 Spector,H.N.
 T  Free-carrier absorption in quasi-two-dimensional
  semiconducting structures
 Phys. Rev. B 28   971-976 (1983)

1575 Srobar,F.
 T  Diffusion in semiconductor superlattices: decay of
  dopant concentration profiles
 Czech. J. Phys. B 33   40-48 (1983)

1576 Stall,R.A.
 Growth of refractory oxide films using solid oxygen
  sources in a molecular beam epitaxy apparatus
 J. Vac. Sci. Technol. B 1   135-137 (1983)

1577 Stanley,C.R., Welch,D., Wicks,G.W., Wood,C.E.C.,
 Palmstrom,C., Pollak,F.H., Parayanthal,P.
 (Al(x)Ga(1-x))(0.47)In(0.53)As; growth and
  characterization
 Inst. Phys. Conf. Ser. 65   173-180 (1983)

1578 Stein,D., Klitzing,K. von, Weimann,G.
 Electron spin resonance on GaAs-al(x)Ga(1-x)As
  heterostructures
 Phys. Rev. Lett. 51   130-133 (1983)

1579 Stern,F.
 T  Doping considerations for heterojunctions
 Appl. Phys. Lett. 43   974-976 (1983)

1580  Stoermer,H.L.
  R   Electron mobilities in modulation-doped GaAs-(AlGa)As
        heterostructures
      Surf. Sci. 132    519-526 (1983)

1581  Stoermer,H.L., Chang,A., Tsui,D.C., Hwang,J.C.M.,
        Gossard,A.C., Wiegmann,W.
      Fractional quantization of the Hall effect
      Phys. Rev. Lett. 50    1953-1956 (1983)

1582  Stoermer,H.L., Haavasoja,T., Narayanamurti,V.,
        Gossard,A.C., Wiegmann,W.
      Observation of the deHaas-van Alphen effect in a
        two-dimensional electron system
      J. Vac. Sci. Technol. B 1    423-426 (1983)

1583  Stoermer,H.L., Schlesinger,Z., Chang,A., Tsui,D.C.,
        Gossard,A.C., Wiegmann,W.
      Energy structure and quantized Hall effect of
        two-dimensional holes
      Phys. Rev. Lett. 51    126-129 (1983)

1584  Stoermer,H.L., Tsui,D.C.
  R   The quantized Hall effect
      Science 220    1241-1246 (1983)

1585  Stoermer,H.L., Tsui,D.C., Gossard,A.C.,
        Hwang,J.C.M.
      Observation of quantized Hall effect and vanishing
        resistance at fractional Landau level occupation
      Physica 117B & 118B    688-690 (1983)

1586  Streda,P., Smrcka,L.
  T   Thermodynamic derivation of the Hall current and the
        thermopower in quantising magnetic field
      J. Phys. C 16    L895-L899 (1983)

1587  Su,S.L., Fischer,R., Lyons,W.G., Tejayadi,O.,
        Arnold,D., Klem,J., Morkoc,H.
      Double heterojunction GaAs/Al(x)Ga(1-x)As bipolar
        transistors prepared by molecular beam epitaxy
      J. Appl. Phys. 54    6725-6731 (1983)

1588  Su,S.L., Lyons,W.G., Tejayadi,O., Fischer,R.,
        Kopp,W., Morkoc,H., McLevige,W.V., Yuan,H.T.
      Molecular beam epitaxial double heterojunction
        bipolar transistors with high current gains
      Electron. Lett. 19    128-129 (1983)

1589  Su,S.L., Tejayadi,O., Drummond,T.J., Fischer,R.,
        Morkoc,H.
      Double heterojunction Al(x)Ga(1-x)As/GaAs bipolar
        transistors (DHBJT's) by MBE with a current gain
        of 1650
      IEEE Electron Device Lett. EDL-4    130-132 (1983)

**1590 Suffczynski,M.**
T  Molecular beam epitaxy and negative ions
    Vacuum 33    369-372 (1983)

**1591 Sugai,S., Harris,J.H., Nurmikko,A.V.**
    Strong electron-phonon interaction effects in modulated transient reflectance spectra of Ga(0.50)In(0.50)P
    Physica 117B & 118B    368-370 (1983)

**1592 Sugimura,A.**
T  Phonon assisted gain coefficient in AlGaAs quantum well lasers
    Appl. Phys. Lett. 43    728-730 (1983)

**1593 Suzuki,Y., Okamoto,H.**
    Refractive index of GaAs-AlAs superlattice grown by MBE
    J. Electron. Mater. 12    397-411 (1983)

**1594 Svensson,S.P., Kanski,J., Andersson,T.G., Nilsson,P.O.**
    Band bending on MBE Al-GaAs(001) as studied by ARUPS
    Surf. Sci. 124    L31-L34 (1983)

**1595 Svensson,S.P., Landgren,G., Andersson,T.G.**
    Al-GaAs(001) Schottky barrier formation
    J. Appl. Phys. 54    4474-4481 (1983)

**1596 Swaminathan,V., Anthony,P.J., Pawlik,J.R., Tsang,W.T.**
    Temperature and excitation dependences of active layer photoluminescence in (Al,Ga)As laser heterostructures
    J. Appl. Phys. 54    2623-2629 (1983)

**1597 Takamori,A., Miyauchi,E., Arimoto,H., Bamba,Y., Hashimoto,H.**
    GaAs molecular beam epitaxy on Be implanted GaAs layers
    Jpn. J. Appl. Phys. 22    L520-L522 (1983)

**1598 Takaoka,H., Chang,C.A., Mendez,E.E., Chang,L.L., Esaki,L.**
    GaSb-AlSb-InAs multi-heterostuctures
    Physica 117B & 118B    741-743 (1983)

**1599 Tan,T.S., Stoneham,E.B., Patterson,G., Collins,D.M.**
    GaAs FET channel structure investigation using MBE
    1983 GaAs IC Symposium Technical Digest    38-41 (1983)

**1600 Tao,R., Thouless,D.J.**
T  Fractional quantization of Hall conductance
    Phys. Rev. B 28    1142-1144 (1983)

1601 Tarucha,S., Horikoshi,Y., Okamoto,H.
Optical absorption characteristics of GaAs-AlGaAs multi-quantum-well heterostructure waveguides
Jpn. J. Appl. Phys. 22   L482-L484 (1983)

1602 Tejayadi,O., Sun,Y.L., Klem,J., Fischer,R., Klein,M.V., Morkoc,H.
Effects of MBE growth conditions on carbon contamination in GaAs
Solid State Commun. 46   251-254 (1983)

1603 Temkin,H., Alavi,K., Wagner,W.R., Pearsall,T.P., Cho,A.Y.
1.5-1.6 um Ga(0.47)In(0.53)As/Al(0.48)In(0.52)As multiquantum well lasers grown by molecular beam epitaxy
Appl. Phys. Lett. 42   845-847 (1983)

1604 Temkin,H., Hwang,J.C.M.
Undoped, semi-insulating GaAs layers grown by molecular beam epitaxy
Appl. Phys. Lett. 42   178-180 (1983)

1605 Thorne,R.E., Su,S.L., Fischer,R., Kopp,W., Lyons,W.G., Miller,P.A., Morkoc,H.
T  Analysis of camel gate FET's (CAMFET's)
IEEE Trans. Electron evices ED-30   212-216 (1983)

1606 Tosatti,E., Parrinello,M.
T  Outline of a theory of the two-dimensional Hall effect in the quantum limit
Lett. Nuovo Cimento 36   289-298 (1983)

1607 Trugman,S.A.
T  Localization, percolation, and the quantum Hall effect
Phys. Rev. B 27   7539-7546 (1983)

1608 Tsang,W.T., Ditzenberger,J.A., Olsson,N.A.
Improvement of photoluminescence of molecular beam epitaxially grown Ga(x)Al(y)In(1-x-y)As by using an As2 molecular beam
IEEE Electron Device Lett. EDL-4   275-277 (1983)

1609 Tsang,W.T., Dixon,M., Dean,B.A.
The characterization and functional reliability of 45 Mbit/s optical transmitters containing MBE-grown lasers
IEEE J. Quantum Electron. QE-19   59-62 (1983)

1610 Tsang,W.T., Hartman,R.L.
cw electro-optical characteristics of graded-index waveguide separate-confinement heterostructure lasers with proton-delineated stripe
Appl. Phys. Lett. 42   551-553 (1983)

1611    Tsang,W.T., Olsson,N.A.
        Preparation of 1.78-um wavelength
            Al(0.2)Ga(0.8)Sb/GaSb double-heterostructure
            lasers by molecular beam epitaxy
        Appl. Phys. Lett. 43    8-10 (1983)

1612    Tsang,W.T., Olsson,N.A.
        New current injection 1.5-um wavelength
            Ga(x)Al(y)In(1-x-y)As/InP double-heterostructure
            laser grown by molecular beam epitaxy
        Appl. Phys. Lett. 42    922-924 (1983)

1613    Tselis,A., Gonzales de la Cruz,G., Quinn,J.J.
  T     Helicon waves in semiconductor superlattices
        Solid State Commun. 47    43-45 (1983)

1614    Tselis,A., Gonzales de la Cruz,G., Quinn,J.J.
  T     Theory of intersubband collective modes in
            semiconductor superlattices
        Solid State Commun. 46    779-781 (1983)

1615    Tsubaki,K., Livingstone,A., Kawashima,M.,
            Okamoto,H., Kumabe,K.
        Differential negative resistance caused by
            inter-subband scattering in a 2-dimensional
            electron gas
        Solid State Commun. 46    517-520 (1983)

1616    Tsui,D.C., Gossard,A.C., Dolan,G.J.
        Radiation effects on modulation-doped
            GaAs-Al(X)Ga(1-x)As heterostructures
        Appl. Phys. Lett. 42    180-182 (1983)

1617    Tsui,D.C., Stoermer,H.L., Hwang,J.C.M.,
            Brooks,J.S., Naughton,M.J.
        Observation of a fractional quantum number
        Phys. Rev. B 28    2274-2275 (1983)

1618    Tu,C.W., Forrest,S.R., Johnston,W.D.,Jr.
        Epitaxial InP/fluoride/InP(001) double
            heterostructure grown by molecular beam epitaxy
        Appl. Phys. Lett. 43    569-571 (1983)

1619    Tu,C.W., Sheng,T.T., Read,M.H., Schlier,A.R.,
            Johnson,J.G., Johnston,W.D.,Jr., Bonner,W.A.
        Growth of single-crystalline epitaxial group II
            fluoride films on InP(001) by molecular beam
            epitaxy
        J. Electrochem. Soc. 130    2081-2087 (1983)

1620    Tzoar,N., Platzman,P.M.
  T     ac conductance of two-dimensional electron gases in a
            semiconductor medium
        Phys. Rev. B 28    4844-4846 (1983)

1621 **Urien,P., Delagebeaudeuf,D.**
T  New method for determining the series resistances in a MESFET or TEGFET
   Electron. Lett. 19   702-703 (1983)

1622 **Valois,A.J., Robinson,G.Y.**
   Characterization of deep levels in modulation-doped AlGaAs/GaAs FET's
   IEEE Electron Device Lett. EDL-4   360-362 (1983)

1623 **Valois,A.J., Robinson,G.Y., Lee,K., Shur,M.S.**
   Temperature dependence of the I-V characteristics of modulation-doped FETs
   J. Vac. Sci. Technol. B 1   190-195 (1983)

1624 **Vasquez,R.P., Lewis,B.F., Grunthaner,F.J.**
   Cleaning chemistry of GaAs(100) and InSb(100) substrates for molecular beam epitaxy
   J. Vac. Sci. Technol. B 1   791-794 (1983)

1625 **Vasquez,R.P., Lewis,B.F., Grunthaner,F.J.**
   Cleaning chemistry of InSb(100) molecular beam epitaxy substrates
   J. Appl. Phys. 54   1365-1368 (1983)

1626 **Vasquez,R.P., Lewis,B.F., Grunthaner,F.J.**
   X-ray photoelectron spectroscopic study of the oxide removal mechanism of GaAs(100) molecular beam epitaxial substrates in 'in situ' heating
   Appl. Phys. Lett. 42   293-295 (1983)

1627 **Vassell,M.O., Lee,J., Lockwood,H.F.**
T  Multibarrier tunneling in Ga(1-x)Al(x)As/GaAs heterostructures
   J. Appl. Phys. 54   5206-5213 (1983)

1628 **Veen,J.F. van der, Smit,L., Larsen,P.K., Neave,J.H.**
   Core level binding energy shifts for reconstructed GaAs(001) surfaces
   Physica 117B & 118B   822-824 (1983)

1629 **Vinter,B.**
T  Subbands in back-gated heterojunctions
   Solid State Commun. 48   151-154 (1983)

1630 **Voisin,P., Bastard,G., Voos,M., Mendez,E.E., Chang,C.A., Chang,L.L., Esaki,L.**
   Optical transmission in GaSb-AlSb superlattices
   J. Vac. Sci. Technol. B 1   409-411 (1983)

1631 Voisin,P., Guldner,Y., Vieren,J.P., Voos,M., Maan,J.C., Delescluse,P., Linh,N.T.
Electron mobility and Landau level width in modulation-doped GaAs-Al(x)Ga(1-x)As heterostructures
Physica 117B & 118B    634-636 (1983)

1632 Wada,O., Miura,S., Ito,M., Fujii,T., Sakurai,T., Hiyamizu,S.
Monolithic integration of a photodiode and a field-effect transistor on a GaAs substrate by molecular beam epitaxy
Appl. Phys. Lett. 42    380-382 (1983)

1633 Wada,O., Sanada,T., Hamaguchi,H., Fujii,T., Hiyamizu,S., Sakurai,T.
AlGaAs/GaAs monolithic LED/amplifier circuit fabricated by molecular beam epitaxy
Jpn. J. Appl. Phys. 22, Suppl. 22-1    587-588 (1983)

1634 Wada,O., Sanada,T., Hamaguchi,H., Fujii,T., Sakurai,T.
Monolithic integration of a double heterostructure light-emitting diode and a field-effect transistor amplifier using molecular beam grown AlGaAs/GaAs
Appl. Phys. Lett. 43    345-347 (1983)

1635 Wada,O., Yamakoshi,S., Sakurai,T., Makiuchi,M., Sanada,T., Miura,S., Hamaguchi,H., Fujii,T., Nakai,K., Nakagami,T.
Monolithic laser/FET and PIN/FET as optoelectronic integrated circuits (OEIC's) by AlGaAs/GaAs heterostructures
1983 GaAs IC Symposium Technical Digest    52-55 (1983)

1636 Wallis,R.H.
Effect of free carrier screening on the electron mobility of GaAs: a study by field-effect measurements
Physica 117B & 118B    756-758 (1983)

1637 Wang,C.S., Grempel,D.R., Prange,R.E.
T  Density-functional theory of a two-dimensional charge-density-wave state in a strong magnetic field
Phys. Rev. B 28    4284-4287 (1983)

1638 Wang,S.Y., Bloom,D.M., Collins,D.M.
20-GHz bandwidth GaAs photodiode
Appl. Phys. Lett. 42    190-192 (1983)

1639 Wang,W.I.
The dependence of Al Schottky barrier height on surface conditions of GaAs and AlAs grown by molecular beam epitaxy
J. Vac. Sci. Technol. B 1    574-580 (1983)

1640 **Wang,W.I.**
Instabilities of (110)III-V compounds grown by molecular beam epitaxy
J. Vac. Sci. Technol. B 1   630-636 (1983)

1641 **Welch,D.F., Wicks,G.W., Eastman,L.F.**
Optical properties of GaInAs/AlInAs single quantum wells
Appl. Phys. Lett. 43   762-764 (1983)

1642 **Welch,D.F., Wicks,G.W., Woodard,D.W., Eastman,L.F.**
GaInAs-AlInAs heterostructures for optical devices grown by MBE
J. Vac. Sci. Technol. B 1   202-204 (1983)

1643 **Whiteaway,J.E.A.**
Simulation and measurements of C/V doping profiles in multilayer structures
IEEE Proc., Part I, 130 No. 4   165-170 (1983)

1644 **Widom,A.**
T  Thermodynamic Hall impedance for two-dimensional models
Phys. Rev. B 28   4858-4859 (1983)

1645 **Williams,G.M., Young,I.M.**
Indium antimonide-strain induced by solder mounting in MBE
J. Cryst. Growth 62   219-224 (1983)

1646 **Wood,C.E.C., Singer,K., Ohashi,T., Dawson,L.R., Noreika,A.J.**
A pragmatic approach to adatom-induced surface reconstruction of III-V compounds
J. Appl. Phys. 54   2732-2737 (1983)

1647 **Wood,C.E.C., Stanley,C.R., Wicks,G.W., Esi,M.B.**
Effect of arsenic dimer/tetramer ratio on stability of III-V compound surfaces grown by molecular beam epitaxy
J. Appl. Phys. 54   1868-1871 (1983)

1648 **Wood,T.H., Burrus,C.A., Miller,D.A.B., Chemla,D.S., Damen,T.C., Gossard,A.C., Wiegmann,W.**
High-speed optical modulation with GaAs/GaAlAs quantum wells in a p-i-n diode structure
IEDM 83 Technical Digest   486-488 (1983)

1649 **Woodall,J.M., Pettit,G.D., Jackson,T.N., Lanza,C., Kavanagh,K.L., Mayer,J.W.**
Fermi-level pinning by misfit dislocations at GaAs interfaces
Phys. Rev. Lett. 51   1783-1786 (1983)

1650  Woodcock,J.M., Harris,J.J.
Control of the height of Schottky barriers on MBE GaAs
Electron. Lett. 19  93-95 (1983)

1651  Woodcock,J.M., Harris,J.J.
Bulk unipolar diodes in MBE GaAs
Electron. Lett. 19  181-183 (1983)

1652  Worlock,J.M., Maciel,A.C., Perry,C.H., Tien,Z.J.,
Aggarwal,R.L., Gossard,A.C., Wiegmann,W.
R  Inelastic light scattering by two dimensional
electron systems in high magnetic fields
Lect. Notes Phys. 177  186-198 (1983)

1653  Xin,S.H., Schaff,W.J., Wood,C.E.C., Eastman,L.F.
Investigations of capless heat treatment of MBE n-GaAs
Inst. Phys. Conf. Ser. 65  613-618 (1983)

1654  Yamada,S., Urisu,T., Mizushima,Y.
T  High-speed infra-red modulator with multilayered
pn-junctions
Electron. Lett. 19  940-941 (1983)

1655  Yamakoshi,S., Sanada,T., Wada,O., Fujii,T.,
Sakurai,T.
AlGaAs/GaAs multiquantum-well (MQW) laser applied to
monolithic integration with FET driver
Electron. Lett. 19  1020-1021 (1983)

1656  Yamanishi,M., Suemune,I.
T  Quantum mechanical size effect modulation light
sources - a new field effect semiconductor laser
or light emitting device
Jpn. J. Appl. Phys. 22  L22-L24 (1983)

1657  Yao,T.
T  A new high-electron mobility monolayer superlattice
Jpn. J. Appl. Phys. 22  L680-L682 (1983)

1658  Yoshida,S., Misawa,S., Gonda,S.
Improvements on the electrical and luminescent
properties of reactive molecular beam epitaxially
grown GaN films using AlN-coated sapphire
substrates
Appl. Phys. Lett. 42  427-429 (1983)

1659  Yoshida,S., Misawa,S., Gonda,S.
Epitaxial growth of GaN/AlN heterostructures
J. Vac. Sci. Technol. B 1  250-253 (1983)

1660  Yoshioka,D.
T  Hall conductivity of two-dimensional electrons in a
periodic potential
Phys. Rev. B 27  3637-3645 (1983)

**1661** Yoshioka,D., Halperin,B.I., Lee,P.A.
T   Ground state of two-dimensional electrons in strong
       magnetic fields and 1/3 quantized Hall effect
    Phys. Rev. Lett. 50   1219-1222 (1983)

**1662** Yoshioka,D., Lee,P.A.
T   Ground-state energy of a two-dimensional
       charge-density-wave state in a strong magnetic
       field
    Phys. Rev. B 27   4986-4996 (1983)

**1663** Yuen,S.Y.
T   Fast relaxing absorptive nonlinear refraction in
       superlattices
    Appl. Phys. Lett. 43   813-815 (1983)

**1664** Zaluzny,M.
T   Interband Raman scattering in size-quantized
       semiconductor films
    Thin Solid Films 100   169-174 (1983)

**1665** Zawadzki,W.
T   Oscillatory magnetization of two-dimensional electron
       gas
    Solid State Commun. 47   317-320 (1983)

**1666** Zeller,C., Vinter,B., Abstreiter,G., Ploog,K.
    Quantization of photo-excited carriers in GaAs doping
       superlattices
    Physica 117B & 118B   729-731 (1983)

**1667** Ziel,A. van der
T   1/f noise in HEMT-type GaAs FETs at low drain bias
    Solid-State Electron. 26   385-386 (1983)

**1668** Ziel,A. van der, Wu,E.N.
T   Thermal noise in high electron mobility transistors
    Solid-State Electron. 26   383-384 (1983)

**1669** Ziel,A. van der, Wu,E.N.
T   High-frequency admittance of high-electron-mobility
       transistors (HEMTs)
    Solid-State Electron. 26   753-754 (1983)

**1670** Ziel,A. van der, Wu,E.N.
T   High-frequency admittance of high-electron-mobility
       transistors (HEMTs)
    Solid-State Electron. 26   753-754 (1983)

**1671** Zipperian,T.E., Dawson,L.R., Osbourn,G.C.,
       Fritz,I.J.
    An In(0.2)Ga(0.8)As/GaAs, modulation-doped,
       strained-layer superlattice field-effect transistor
    IEDM 83 Technical Digest   696-699 (1983)

**1672** Zucker,J.E., Pinczuk,A., Chemla,D.S., Gossard,A.C., Wiegmann,W.
Raman scattering resonant with quasi-two-dimensional excitons in semiconductor quantum wells
Phys. Rev. Lett. 51    1293-1296 (1983)

**1673** Zurcher,P., Bauer,R.S.
Photoemission determination of dipole layer and VB-discontinuity formation during the MBE growth of GaAs on Ge(110)
J. Vac. Sci. Technol. A 1    695-700 (1983)

# Author Index

Abdalla,M.I.  302,388
Abe,M.  644,744,745,1173,
  1174,1486
Abram,R.A.  1228
Abstreiter,G.  216,289,386,
  509,564,746,1169,1170,
  1666
Adachi,S.  747
Afsar,M.N.  510
Aggarwal,R.  1123
Aggarwal,R.L.  549,1124,
  1652
Aishima,A.  1175
Alavi,K.  817,1049,1110,
  1111,1176,1177,1178,
  1240,1260,1263,1269,
  1462,1515,1516,1570,1603
Aleksandrov,L.N.  387
Alexandre,F.  302,388,572,
  748,749
Alferov,Z.I.  511
Allen,S.J.,Jr.  1179
Allyn,C.L.  389,899,1312,
  1395
Almassy,R.J.  311
Altarelli,M.  686,986,1180,
  1181,1182
Alvarado,S.F.  512,513,514,
  750,835,1236
Anderson,P.W.  1183
Anderson,R.J.  1194,1423,
  1479
Andersson,T.G.  515,516,
  751,752,753,973,1046,
  1184,1185,1594,1595
Ando,S.  468,656
Ando,T.  290,460,461,517,
  518,519,754,755,756,757,
  758,760,1186
Andreoni,W.  217,291
Andrews,D.A.  559,1187,1366

Anh,V.H.  1564
Ankri,D.  759,1188,1189,
  1190
Anthony,P.J.  1191,1596
Aoki,H.  519,760,1192
Aono,M.  1264
Appelbaum,J.A.  158
Arakawa,Y.  761
Arimoto,H.  1597
Arnold,D.  1293,1294,1314,
  1398,1451,1587
Arora,V.K.  1427
Arthur,J.R.,Jr.  4,5,6,8,
  38,53,83,292
Asahi,H.  293,340,520,521,
  607,608,762,763,764,950,
  951,1193,1389,1390,1391
Asatourian,R.  765
Asbeck,P.M.  522,765,766,
  1015,1016,1194,1195,
  1351,1479
Ashenford,D.  1095
AuCoin,T.R.  444,445,633
Auvray,P.  1211

Babcock,E.J.  1195
Baceiredo,S.  1038
Bachmann,K.J.  156,188
Bachrach,R.Z.  294,295,523,
  524,525,767,1196,1197,
  1231,1407
Bafleur,M.  768,769,1198,
  1199
Bajaj,K.K.  1341,1342,1343,
  1517
Baldereschi,A.  217,291
Baldwin,K.  699
Balibar,F.  157
Balk,P.  723
Ballamy,W.C.  84,116

Ballingall,J.M.  526,770,
    1200,1201,1202
Ballini,Y.  471
Ballmann,A.A.  814
Bamba,Y.  1203,1597
Bandy,S.G.  296
Banerjee,S.  611
Baraff,G.A.  158,527
Barker,A.S.,Jr.  189,218
Barnard,J.A.  390,469,470,
    528,529,657,771,772,
    1051,1188
Barnes,P.A.  78,219
Barnett,S.A.  739
Barret,C.  1204,1205
Barrett,J.H.  773,774,1206
Bass,F.G.  391
Bastard,G.  530,531,724,
    775,776,777,778,1008,
    1207,1208,1209,1467,1630
Basu,B.K.  779
Basu,P.K.  392,532,1210
Bau,N.K.  274
Baudet,M.  1211
Bauer,R.S.  295,523,524,
    780,781,1212,1213,1673
Bauser,E.  216
Bayraktaroglu,B.  1214,
    1215,1561
Bean,J.C.  297
Beard,W.T.  1010
Beeby,J.L.  782
Beerens,J.  880
Beneking,H.  783
Benoit a la Guillaume,C.
    557
Benoit,C.  503
Ber,B.Y.  511,645
Berenz,J.  614
Berezhkovskil,A.M.  784
Beril,S.I.  1525
Berry,J.D.  507,647,1511
Berson,B.  298
Bethea,C.G.  550,551,785,
    816,818,821,978,1216,
    1240,1241,1242,1243,
    1261,1433,1434
Bhattacharya,B.K.  1217
Bhattacharya,K.  779
Bhattacharya,P.K.  786,787
Bhattacharyya,K.  1210
Bianconi,A.  295
Biefeld,R.M.  1322
Bischoff J.C.  1500
Blakemore,J.S  1283

Blakey,P.A.  1218
Blanchet,R.C.  227,299,393,
    533,788
Bliek,L.  1219,1220
Blood,P.  394,465,789,790,
    1221
Bloom,D.M.  1146,1638
Bloss,W.L.  791,792,793,
    1222,1223,1224,1225
Bluyssen,H.J.A.  300,794
Board,K.  370,416,444,445,
    534,574,633,648,696,
    1026,1156
Bok,J.  1226
Bolmont,D.  824
Bonner,W.A.  552,555,1131,
    1619
Bonnevie,D.  795
Borovitskaya,E.S.  1227
Bosch,R.  425,426
Bosman,G.  1556
Bouchaib,P.  535
Boudry,M.R.  68
Bovin,V.P.  265
Brand,S.  1228
Braun,E.  1219,1220,1229
Braunstein,M.  885
Brenac,A.  302,388,536,537
Brenig,W.  1230
Brennan,T.M.  1372,1373
Bringans,R.D.  767,1196,
    1197,1231
Brinkman,W.F.  1455
Briones,F.  796
Brody,E.M.  792
Brooks,J.S.  1617
Brown,W.  899
Brummell,M.A.  1072,1232,
    1495,1496
Budarnykh,V.I.  395
Buehler,E.  156
Buehlmann,H.J.  786,787,
    1500
Burkhard,H.  729
Burnham,R.D.  205,224,301
Burroughs,M.S.  598
Burrus,C.A.  440,441,1648
Busserand,B.  1233
Button,K.J.  510
Bye,K.L.  394


Cadoret,R.  797
Cage,M.E.  1134,1234,1235,
    1329

Calawa,A.R.   220,320,408,
    538,982,1461,1471,1472
Caldwell,P.J.   1414
Calecki,D.   1320
Cameron,D.C.   580
Camp,W.O.   629
Campagna,M.   512,513,750,
    1236
Camras,M.D.   597,598,1030,
    1035
Capasso,F.   539,798,799,
    800,801,802,803,804,841,
    978,1131,1150,1237,1238,
    1239,1240,1241,1242,
    1243,1244,1352,1353,
    1354,1433,1462
Car,R.   217,291
Carelli,J.   945
Carenco,A.   302
Caridi,E.A.   814,1056,1504,
    1565,1566
Carlsson,A.E.   1245
Carter,C.B.   540,805
Caruthers,E.   159,160
Casey,H.C.,Jr.   56,57,78,
    79,94,121,166,221,222,
    239,303,304,305,403
Castano,J.L.   1199
Ch'en,D.R.   120
Chai,Y.G.   396,443,541,542,
    806,997,1246,1270,1271
Chaikovskii,I.A.   146,161,
    208,274,543
Chaix,C.   1277
Chakraborti,N.B.   1531
Chakravarti,A.N.   1326
Chalker,J.T.   1247
Chandra,A.   534
Chang,A.   1581,1583
Chang,A.M.   1248
Chang,C.A.   136,162,164,
    202,266,270,306,397,433,
    451,484,544,545,546,547,
    548,692,807,808,809,810,
    811,834,911,1249,1250,
    1251,1272,1408,1468,
    1598,1630
Chang,L.L.   32,39,40,41,45,
    46,54,55,66,73,80,110,
    117,118,124,125,136,162,
    163,164,172,202,266,270,
    300,307,308,318,365,397,
    398,399,400,423,433,451,
    503,544,547,548,549,576,
    631,636,724,777,778,794,
    811,812,813,905,986,988,
    1008,1009,1208,1209,
    1252,1467,1468,1469,
    1530,1598,1630
Chang,R.P.H.   376
Chang,T.Y.   814,1056,1504,
    1565,1566
Chang,Y.C.   687,815,990,
    991,992,1253,1254
Chao,P.C.   606
Chaplart,J.   349,409,446,
    562,627,634,848,849,977,
    979,1137,1138,1286
Chaplik,A.V.   401,1141
Chapman,R.L.   320
Chattopadhyay,D.   1255,1256
Chaudhuri,S.   1257
Chekir,F.   1204
Chemla,D.S.   1014,1258,
    1259,1477,1478,1648,1672
Chen,C.Y.   550,551,785,816,
    817,818,819,820,821,822,
    823,1216,1260,1261,1262,
    1263,1512
Chen,P.   824
Cheng,H.   1436
Cheng,K.Y.   441,552,553,
    554,555,556,819,820,822,
    825,826,827,828,869,946,
    1049,1072,1232,1385,
    1495,1496
Cherevatskii,N.Y.   331,415
Chevrier,J.   1003,1286,1539
Chew,N.G.   843
Chiang,T.C.   829,984,1264,
    1449
Chiaradia,P.   523,524
Chien,W.Y.   239
Chin,M.A.   355
Chinen,K.   402,1024
Chiu,L.C.   1265,1266,1267,
    1571
Cho,A.Y.   9,10,11,14,15,16,
    17,18,19,20,21,28,29,56,
    57,58,59,78,81,82,83,84,
    115,116,119,120,121,133,
    134,140,165,166,167,168,
    169,199,203,209,219,221,
    222,223,239,240,241,242,
    272,280,286,303,304,305,
    309,314,348,353,403,404,
    440,441,452,453,462,463,
    464,474,475,501,510,550,
    551,552,553,554,555,556,
    560,565,567,568,569,570,

Cho, A.Y.
  628, 666, 689, 738, 739, 783,
  785, 816, 817, 818, 819, 820,
  821, 822, 823, 825, 826, 827,
  828, 830, 831, 854, 860, 864,
  865, 869, 870, 946, 962, 964,
  982, 1032, 1049, 1065, 1072,
  1110, 1111, 1145, 1176,
  1177, 1178, 1216, 1232,
  1240, 1260, 1261, 1262,
  1263, 1267, 1268, 1269,
  1318, 1372, 1385, 1397,
  1462, 1495, 1496, 1515,
  1516, 1557, 1570, 1571, 1603
Chomette, A.   832, 1063
Choudhury, A.N.M.M.   1110,
  1111, 1269, 1570
Chow, R.   541, 542, 806, 1270,
  1271
Christman, S.B.   828
Christou, A.   833
Chu, W.K.   484, 544, 834, 1250,
  1272, 1521
Churchill, J.N.   1273
Ciccacci, F.   512, 513, 835,
  1236
Clampitt, R.   663
Clegg, J.B.   604, 836
Cohen, M.J.   1011
Cohen, M.L.   336
Cohen, P.I.   926, 1367, 1368,
  1369
Coleman, J.J.   508
Coleman, P.D.   837, 838
Collins, D.M.   296, 310, 334,
  345, 796, 839, 840, 983,
  1071, 1094, 1146, 1200,
  1274, 1275, 1276, 1301,
  1302, 1313, 1599, 1638
Colvard, C.   405, 452, 453,
  1317
Comas, J.   250, 406
Combescot, M.   557, 1226
Contour, J.P.   535, 1277
Conwell, B.M.   224
Cook, R.K.   1278
Cooper, J.A., Jr.   841
Covington, D.W.   225, 226,
  311, 312, 406, 833
Cox, N.W.   842, 921
Crowell, C.R.   1432
Cullis, A.G.   230, 313, 321,
  418, 843

D'Avitaya, F.A.   535

D'Haenens, I.J.   885
Daembkes, H.   1446
Daeweritz, L.   844
Dalman, G.C.   614
Damen, T.C.   1259, 1648
Dandekar, N.   416, 431, 574,
  696
Dandekar, N.V.   346
Das Sarma, S.   407, 558, 845,
  846, 1279, 1280
Datta, S.   1281
Davey, J.E.   2, 7, 833
Davies, G.J.   408, 427, 428,
  559, 1187, 1355, 1366
Dawson, L.R.   1322, 1323,
  1324, 1521, 1555, 1646, 1671
Dawson, M.D.   1282
Dawson, P.   683, 690, 874, 887,
  1088, 1282, 1490
Day, D.S.   560
DeFreez, R.K.   1283
DeJong, T.   1284, 1285
DeSimone, D.   506, 742, 805,
  847, 1155, 1163
Dean, B.A.   1609
Dekkers, J.J.M.   783
Delagebeaudeuf, D.   409, 561,
  562, 627, 848, 849, 850, 851,
  852, 977, 979, 980, 1137,
  1138, 1286, 1539, 1621
Delescluse, P.   409, 562, 627,
  848, 849, 852, 977, 979, 980,
  998, 999, 1003, 1137, 1138,
  1286, 1539, 1631
Delgado, J.   632
Delhomme, B.J.   227, 299, 393,
  533, 788
Deline, V.   1356
Dernier, P.D.   223
Deveaud, B.   971, 1416
Devlin, W.J.   410, 696
Devoldere, P.   186, 246, 350
Dezaly, F.   447, 448
DiLorenzo, J.V.   167, 168,
  314, 853, 927
Dingle, R.   60, 85, 86, 87, 88,
  95, 96, 113, 122, 129, 141,
  228, 275, 297, 315, 358, 359,
  371, 372, 377, 411, 412, 477,
  478, 479, 667, 731, 732, 733,
  734, 853, 854, 946, 1312,
  1385, 1395
Ditzenberger, J.A.   500, 715,
  1130, 1132, 1133, 1608
Dixon, M.   1609

Dixon,R.W.   121
Dixon,S.   996
Djafari-Rouhani,B.   1287,
  1551
Dobrzynski,L.   1287,1551
Dobson,P.J.   588,855,856,
  1088,1417,1490,1491,1493
Doehler,G.H.   30,31,89,111,
  261,316,317,413,563,564,
  618,674,676,746,857,858,
  943,968,1288,1289,1290,
  1291,1383,1384,1523,
  1535,1536,1542,1543,
  1544,1545
Doerbeck,F.H.   1007
Dolan,G.J.   1616
Dordzhin,G.S.   859
Dorfman,V.F.   414
Douma,W.A.S.   1284,1285
Dove,D.B.   41
Dow,J.D.   646
Dowsett,M.G.   170
Driscoll,P.   967
Drummond,T.J.   464,504,560,
  565,566,567,568,569,570,
  597,598,610,660,697,739,
  860,861,862,863,864,865,
  866,867,868,869,870,871,
  872,873,922,923,954,955,
  961,962,963,964,1007,
  1030,1031,1032,1033,
  1034,1035,1036,1082,
  1100,1118,1119,1120,
  1292,1293,1294,1314,
  1315,1316,1317,1318,
  1397,1398,1428,1429,
  1430,1431,1589
Duggan,G.   571,690,691,874,
  887,1088,1282
Duh,K.H.   1295
Duhamel,N.   572,748
Duncan,W.M.   1007,1561
Dunn,C.N.   58
Dupas,G.   1211
Dutta,N.K.   1296
Dvoryankin,V.F.   331,415
Dvoryankina,G.G.   331,415
Dziuba,R.F.   1134,1234

East,J.R.   1218
Eastman,D.E.   829,984,1264,
  1449
Eastman,L.F.   369,370,390,
  410,416,431,444,445,454,
  455,469,470,487,488,490,
  507,526,528,529,534,573,
  574,606,629,633,637,647,
  648,649,657,658,659,696,
  727,728,737,759,771,772,
  875,876,877,949,983,
  1026,1054,1055,1156,
  1162,1163,1188,1189,
  1190,1202,1297,1365,
  1370,1371,1461,1510,
  1511,1514,1556,1641,
  1642,1653
Ebert,G.   878,1220,1298,
  1299,1400
Eden,R.C.   1300
Edlinger,J.   1440
Eglash,S.J.   1094,1274,
  1301,1302
Ehrenreich,H.   1245
Eilenberger,D.J.   1014,1477
Eisele,F.L.   921,1364
Eisen,F.H.   1194
Eisen,P.H.   1479
Eisenberger,P.   348
Ekenberg,U.   1303
Elachi,C.   123
Elder,H.E.   498
Elliott,R.A.   1283
Elta,M.E.   1218
Enaki,N.A.   543,1091
Engelmann,H.J.   1219
Englert,T.   501,879,880,
  987,1304,1305,1306,1307,
  1308,1453
Ennen,H.   1093
Epshtein,E.M.   144,145,195,
  208
Erickson,L.P.   655,1047,
  1294,1309,1310
Esaki,L.   12,27,32,39,40,
  45,46,52,54,55,61,62,63,
  64,65,66,72,73,89,100,
  110,112,118,124,125,136,
  148,162,163,164,171,172,
  201,202,206,279,266,267,
  268,269,270,300,307,308,
  318,319,365,397,400,417,
  423,433,451,484,503,544,
  547,548,549,575,576,631,
  636,724,725,777,778,794,
  811,813,834,881,882,905,
  986,988,1008,1009,1208,
  1209,1467,1468,1469,
  1598,1630
Esi,M.B.   1647
Etienne,P.   137,173,186,
  351,409,448,848,852,977,
  998,1003,1277

Eu.V.K.  885
Evans,C.A.,Jr.  463,847,
    868,1317,1565,1566

Fabre,N.  651
Fan,J.C.C.  320
Farrow,R.F.C.  67,69,174,
    230,231,313,321,418,577,
    578,579,580,702,843,883,
    1106,1311
Favennec,P.N.  884
Feder,R.  513,514
Fedirko,V.A.  276
Feldman,L.C.  666,738,1518,
    1519,1520
Feng,M.  885
Ferry,D.K.  581,904,929,
    1078,1079,1080,1537
Feuer,M.  853
Feuer,M.D.  1312,1395
Field,B.F.  1134,1234
Fischer,A.  197,198,216,
    236,262,263,362,363,384,
    438,618,619,673,674,675,
    676,886,1409,1410,1453
Fischer,B.  1275,1276,1313
Fischer,R.  570,860,861,
    863,864,867,868,871,872,
    962,964,1008,1031,1032,
    1033,1035,1036,1100,
    1101,1118,1119,1120,
    1292,1293,1294,1309,
    1314,1315,1316,1317,
    1318,1397,1398,1431,
    1470,1573,1587,1588,
    1589,1602,1605
Fishman,G.  1319,1320
Flahive,P.G.  928
Fleming,R.M.  419
Flemming,K.E.  1046,1184
Fonstad,C.G.  1110,1111,
    1269,1570
Fork,R.L.  1090,1560
Forrest,S.R.  1176,1618
Fowler,A.B.  757
Foxon,C.T.  42,43,68,69,90,
    97,126,175,181,182,183,
    232,233,238,420,571,582,
    583,604,736,836,874,887,
    1095,1282,1321
Foy,P.W.  166,303,304,305,
    403,801,1240
Frahm,R.E.  1373
Fraley,P.E.  717,718,719

Francombe,M.H.  654,1044,
    1045,1499
Freeman,J.  838
Freeouf,J.L.  612,613,743,
    1501
Freller,H.  888
Frensley,W.R.  1007
Friedel,P.  1337
Friedman,L.  793,1225
Fritz,I.J.  1322,1323,1324,
    1555,1671
Fujii,T.  178,322,335,429,
    430,458,592,593,595,596,
    889,1102,1164,1165,1325,
    1632,1633,1634,1635,1655
Fujii,Y.  385
Fukai,M.  192
Fukushima,Y.  1175
Fukuyama,H.  890
Fulco,P.  896
Fulton,R.C.  1433
Fumey,M.  227
Furuya,T.  1203

Galuska,A.  1511
Ganser,P.M.  489,1093
Gant,H.  459,891,1040
Gaponov,S.V.  584
Garbinski,P.A.  550,551,
    785,816,817,818,819,820,
    821,822,823,1216,1260,
    1261,1262,1263,1512
Garner,C.M.  323
Gaucherel,P.  1513
Gauneau,M.  1211
Geng,P.  127
Genkin,V.M.  1227
Ghatak,K.P.  1326
Ghibaudo,G.  892
Ghosal,A.  1256
Gibbs,H.M.  324,325,326,
    327,476,893,894,939,
    1116,1327,1382
Gibson,J.M.  1328,1518,
    1519,1520
Girvin,S.M.  682,1234,1235,
    1329
Giuliani,G.F.  1330,1331,
    1529
Glasser,M.L.  1332
Glisson,T.H.  421,1443
Gmelin,E.  1171
Goebel,E.O.  1333,1383
Goldstein,L.  895
Golio,J.M.  1444

Goncalves da Silva,C.E.T.
  724,896
Gonda,S.  91,128,138,139,
  234,249,385,1167,1168,
  1658,1659
Gonzales de la Cruz,G.
  1334,1613,1614
Goodhue,W.D.  1572
Gornik,E.  422,585,920,
  1363,1419,1440
Gossard,A.C.  87,102,129,
  133,141,189,190,214,215,
  218,228,259,275,287,288,
  315,324,325,326,327,328,
  352,355,357,358,359,371,
  372,389,405,411,412,419,
  422,453,457,476,477,478,
  479,501,502,585,640,642,
  667,669,698,699,700,721,
  722,731,732,733,734,823,
  853,879,880,893,894,897,
  898,899,900,901,909,910,
  920,939,987,1014,1019,
  1020,1021,1057,1058,
  1062,1065,1066,1090,
  1096,1097,1116,1124,
  1134,1135,1136,1234,
  1259,1305,1306,1307,
  1308,1327,1335,1352,
  1353,1354,1363,1382,
  1477,1478,1480,1481,
  1482,1505,1559,1560,
  1581,1582,1583,1585,
  1616,1648,1652,1672
Gotoh,H.  586,902,1336
Gourley,P.L.  1322
Gourrier,S.  1337
Gover,A.  92
Gowers,J.P.  684,908,1158,
  1338,1492
Graf,K.  619,886
Grange,J.D.  329,330,450,
  664,903
Grant,A.J.  230,321
Grant,R.W.  616,726,965,
  1339,1554
Grassie,A.D.C.  1221
Greene,B.I.  1090,1560
Greene,J.E.  739,1082,1340
Greene,R.L.  1341,1342,1343
Greggi,J.,Jr.  1499
Grempel,D.R.  1637
Griem,T.  805
Griffiths,G.  1406
Griffiths,R.M.  69

Grondin,R.O.  904,1079,
  1080,1537
Grubin,H.L.  1344
Grunthaner,F.J.  1345,1624,
  1625,1626
Guenther,K.G.  1,3,888
Guldner,Y.  423,503,631,
  905,1346,1631
Gulyaev,Y.V.  331
Gunshor,R.L.  1281

Haavasoja,T.  1582
Haddad,G.I.  1218
Hafendoerfer,M.  619,886
Hagino,M.  402,1024
Haisma,J.  1284,1285
Hajdu,J.  1347
Haldane,F.D.M.  1348
Halperin,B.I.  906,1349,
  1532,1661
Hamaguchi,H.  1633,1634,
  1635
Hamann,D.R.  158
Hansson,G.V.  523,524
Hardouin-Duparc,O.  1287
Harland,C.J.  150
Harris,J.H.  907,1103,1591
Harris,J.J.  380,424,571,
  587,588,789,908,1350,
  1490,1650,1651
Harris,J.S.,Jr.  456,508,
  522,1016,1018,1351
Harrison,W.A.  176,267
Harrold,C.J.  814
Hartman,R.L.  121,498,665,
  716,717,1296,1610
Harvey,J.A.  43
Hashimoto,H.  430,592,593,
  595,596,600,603,652,653,
  889,931,960,1039,1042,
  1325,1403,1597
Hass,K.C.  1245
Hata,T.  1050,1502
Hauser,J.R.  421,1443,1444
Hayakawa,H.  385
Hayashi,I.  17,18,19,20
Haydl,W.H.  425,426,589
Hayes,J.R.  539,1352,1353,
  1354,1462
Heckingbottom,R.  408,427,
  428,559,1187,1355
Hegarty,J.  909,910
Heiblum,M.  545,546,911,
  1043,1356,1469,1470
Heime,K.  1446

## Author Index

Heinonen, O. 1357
Hendel, R. 853, 1515
Hendel, R.H. 1312, 1395
Henoc, P. 572, 895
Henry, C.H. 60
Henry, L. 884
Herman, F. 235
Herman, M.A. 912, 1358
Hernandez-Calderon, I. 1359, 1362
Herrick, D.R. 177
Herzog, T. 1401
Hess, K. 332, 333, 421, 565, 567, 569, 590, 597, 598, 610, 611, 646, 838, 922, 923, 954, 955, 1303, 1360, 1374, 1422, 1443
Hesto, P. 1361
Hewitt, B.S. 167
Hicklin, W.H. 225
Hickmott, T.W. 1573
Hida, H. 600, 931, 1375
Hierl, T.L. 334, 614, 913
Hiesinger, P. 489, 615
Higgins, J.A. 1473
Hikosaka, K. 591, 643, 644, 914
Himpsel, F.J. 1264
Hirofuji, Y. 139
Hiroi, S. 1004, 1005
Hirose, M. 236, 363, 383, 384
Hisatsugu, T. 335
Hitzman, C.J. 1510, 1565, 1566
Hiyamizu, S. 178, 322, 335, 429, 430, 458, 592, 593, 594, 595, 596, 600, 602, 603, 643, 644, 652, 653, 889, 914, 915, 916, 917, 918, 931, 936, 960, 1039, 1042, 1102, 1164, 1165, 1325, 1375, 1376, 1377, 1403, 1488, 1498, 1547, 1632, 1633
Hjalmarson, H.P. 919
Hoechst, H. 1359, 1362
Hoepfel, R.A. 920, 1363
Holah, G.D. 921, 1364
Holbrook, W.R. 498, 717, 718, 719
Holden, W.S. 240, 628
Hollan, L. 130
Hollis, M. 431, 574, 1370, 1510, 1556
Holloway, S. 126, 132
Holm-Kennedy, J.W. 281

Holmstrom, F.E. 1273
Holonyak, N., Jr. 597, 598, 922, 923, 1030, 1035
Hong, C.S. 1365
Hooft, G.W. t' 874, 887
Hopkins, C.G. 463, 868, 1317
Hopster, H. 514, 750
Horikoshi, Y. 1379, 1380, 1503, 1601
Hoskins, M.J. 924
Hotta, T. 925, 1089
Hottmann, H. 70
Hou, D. 1423
Houghton, A.J.N. 1366
Hove, J.M. van 926, 1367, 1368, 1369
Howard, W.E. 32, 39, 40
Hsieh, K.H. 1370, 1371
Huet, D. 795
Hulyer, P.J. 1095
Hung, C.Q. 161
Hunsinger, B.J. 924
Hutchinson, A.L. 801, 802, 803, 1242, 1243
Hwang, J.C.M. 853, 927, 928, 1049, 1179, 1248, 1312, 1372, 1373, 1395, 1509, 1581, 1585, 1604, 1617

Iafrate, G.J. 929, 1079, 1080, 1374
Igarashi, O. 234
Ignatov, A.A. 93, 131
Ihm, J. 336, 599
Ikeda, M. 293, 340, 520, 521, 607
Ikegami, T. 763, 764, 950, 1391
Ilegems, M. 79, 94, 95, 96, 114, 133, 152, 153, 154, 179, 210, 237, 250, 264, 375, 786, 787, 1500
Imry, Y. 930
Inada, M. 1160
Inada, T. 1498
Inayama, M. 1375, 1376, 1377
Inoue, M. 600, 931, 1375, 1376, 1377
Inuishi, Y. 600, 931, 1375, 1376, 1377
Inuzuka, H. 947, 948
Iordansky, S.V. 932
Ishibashi, T. 601, 933, 934, 935, 937, 1117, 1379
Ishii, T. 285

Ishikawa,H.   602,744,745
Ishikawa,T.   603,889,918,
   936,960,1403,1498,1547
Isihara,A.   1378
Ito,M.   1632
Itoh,A.   385
Itoh,S.   254
Ivanov,I.   337
Iwakuro,H.   1412
Iwamura,H.   937,1379,1380,
   1402

Jackson,T.N.   743,1489,1649
Jacobi,K.   127,338,437,681,
   1381
Jaekel,M.T.   1532
Janak,J.   77
Janssen,A.P.   150,180
Janssen,G.   183
Jenkinson,H.A.   938
Jewell,J.L.   893,894,939,
   1116,1327,1382
Jewsbury,P.   132
Jha,S.S.   37
Jiang,D.S.   940
Joannopoulos,J.D.   599
Johannessen,J.S.   604
Johnson,J.G.   1619
Johnson,W.C.   292
Johnston,W.D.,Jr.   1618,
   1619
Jones,G.R.   321,418,579,
   580,702,1106
Joshin,K.   591,643,644,1485
Joyce,B.A.   43,68,69,71,90,
   97,126,150,175,181,182,
   183,233,238,256,257,284,
   339,342,420,465,583,587,
   588,604,605,624,736,789,
   855,856,908,941,942,975,
   1140,1158,1417,1490,
   1491,1492,1493
Joynt,R.   1073
Judaprawira,S.   506,534,
   606,727
Jung,H.   943,944,969,970,
   1333,1383,1384,1524

Kaelin,G.R.   522
Kahn,A.   945
Kakati,D.   98
Kamarinos,G.   892
Kamei,H.   1113
Kamimura,K.   106
Kaminsky,G.   722

Kanski,J.   1594
Kasamanyan,Z.H.   432
Kasowski,R.V.   235
Kastalsky,A.   853,946,1385,
   1386
Kaushik,S.B.   1387
Kavanagh,K.L.   1649
Kawabe,M.   947,948,1388
Kawai,N.J.   308,433,503,
   548,813,949,1487,1502
Kawamura,Y.   293,340,341,
   520,521,607,608,762,763,
   764,950,951,1193,1389,
   1390,1391
Kawanami,H.   952
Kawashima,M.   747,1112,
   1445,1487,1615
Kazarinov,R.F.   22,23,33,
   44,609,900,953,1450
Keever,M.   565,610,611,838,
   862,865,866,954,955
Kelly,M.J.   1392
Keramidas,V.G.   853,1312,
   1395
Kervarec,J.   1233,1552
Khapachev,Y.P.   1393,1394
Kido,G.   956
Kiehl,R.A.   853,1312,1395
Kim,J.Y.   957
Kim,K.   1414
Kimata,M.   139,213,285,381,
   382,586,902,1336
King,R.M.   170,251,330,450,
   664,1064
Kirchner,P.D.   490,612,613,
   743,1489
Kirkpatrick,C.G.   765,766,
   1195
Kitahara,K.   958
Klein,M.V.   405,452,453,
   1602
Kleinman,D.A.   352,457,639,
   640,1021,1376,1433
Klem,J.   1293,1317,1318,
   1397,1398,1470,1587,1602
Klitzing,K. von   878,959,
   1298,1299,1399,1400,
   1401,1578
Knecht,J.   676,970,1409,
   1410
Knodle,W.S.   434,443,927
Kobayashi,H.   1380,1402
Kobayashi,M.   1173,1486
Kocot,C.   1275,1276
Kogelnik,H.   435

Kohmoto,M.   1122
Kollberg,E.   831,1557
Koma,A.   112,117,135
Kometani,T.Y.   99
Konagai,M.   373,1004,1005
Kondo,K.   960,1403
Kondo,M.   1388
Kondoh,H.   614
Kop'ev,P.S.   511,645
Kopp,W.   565,566,567,863,
   864,865,866,867,871,872,
   873,955,961,962,963,964,
   1033,1036,1100,1119,
   1120,1294,1315,1318,
   1451,1588,1605
Koschel,W.H.   489,615,701
Kose,V.   1229
Koszi,L.A.   237
Kowalczyk,S.P.   616,726,
   945,965,1151,1339,1554
Krasheninnikov,M.V.   401
Kraut,E.A.   726,965,1339
Kreskovsky,J.P.   1344
Krikorian,E.   1037
Krishna,P.   206
Kroemer,H.   239,436,617,
   966,1160,1161,1404,1405,
   1406
Krusor,B.S.   525,1407
Ktitorov,S.A.   24
Kuan,T.S.   1251,1408
Kubiak,R.A.   664,967
Kuebler,B.   437
Kuenzel,H.   362,438,439,
   564,618,619,620,674,675,
   676,746,858,886,943,944,
   968,969,970,1070,1071,
   1384,1409,1410,1453,
   1524,1535
Kuhl,J.   1333
Kumabe,K.   449,747,1445,
   1615
Kunc,K.   621
Kuramoto,K.   1203
Kuramoto,Y.   1411
Kuroda,S.   1336,1485
Kuroda,T.   1412
Kuvas,R.L.   58
Kuznetsov,G.F.   1394
Kwapien,W.M.   1443


Lai,W.Y.   1413
Laidig,W.D.   597,598,938,
   1414,1415
Lam,P.K.   336
Lambert,B.   971,1416
Landgren,G.   622,636,752,
   753,972,973,1185,1264,
   1595
Landwehr,G.   623
Lang,D.V.   133,221,305,635,
   1006
Lanza,C.   1649
Larsen,P.K.   342,624,625,
   974,975,1140,1337,1417,
   1418,1492,1493,1628
Lassnig,R.   1419,1420,1440
Laughlin,R.B.   626,976,1421
Laurence,G.   183,343
Laviron,M.   409,562,627,
   634,848,849,852,977,979,
   980,1137,1138,1286,1539
LePore,J.J.   8
Leburton,J.P.   1422
Lebwohl,P.A.   13
Lee,C.A.   614
Lee,C.P.   1423,1424,1432
Lee,C.S.   323
Lee,J.   1425,1426,1427,1627
Lee,J.W.   1414,1415
Lee,K.   871,872,1292,1315,
   1398,1428,1429,1430,
   1431,1623
Lee,P.A.   1661,1662
Lee,R.F.   486
Lee,S.J.   1423,1432
Lee,T.P.   134,240,440,441,
   628
Leheny,R.F.   207,814
Lemarchand,A.   1142
Lent,C.S.   1367,1368,1369
Leontiew,H.   1219
Leroux,M.   1277
Levesque,B.   1277
Levine,B.F.   551,978,1216,
   1241,1242,1243,1261,
   1433,1434
Levine,H.   1435
Levy,H.M.   454,629,637
Lewis,B.F.   1624,1625,1626
Ley,L.   244,344
Li,A.Z.   1436,1437,1438,
   1439
Libby,S.B.   1435
Lin-Chung,P.J.   159,160
Lind,M.D.   1554
Lindau,I.   693,694,1094
Lindemann,G.   920,1440

Linh,N.T.   137,173,186,246,
    349,350,351,409,446,447,
    448,561,562,627,634,848,
    849,850,851,852,977,979,
    980,998,999,1000,1001,
    1002,1003,1137,1138,
    1286,1441,1442,1539,1631
Linke,R.A.   203,241,242
Littlejohn,M.A.   421,938,
    1415,1443,1444
Litton,C.W.   311
Liu,Y.Z.   981
Livingston,A.R.   1300
Livingstone,A.   1445,1615
Lockwood,H.F.   1627
Logan,R.A.   51,102,141,275,
    419,499,539,800,1066,
    1129,1130,1434,1540
Logvinskiy,L.M.   395
Loreck,L.   1446
Low,T.S.   982,983,1447
Ludeke,R.   39,40,41,45,72,
    73,80,100,135,162,202,
    243,244,308,344,442,484,
    545,622,636,829,834,972,
    984,1264,1448,1449
Ludowise,M.   954
Lueth,H.   723
Lugli,P.   904
Lunardi,L.   1188
Luo,W.   652,653,1042
Luryi,S.   609,900,953,1241,
    1386,1450
Luscher,P.E.   184,345,434,
    443,630,913,927
Luskin,B.M.   584
Lykakh,V.A.   391,985
Lyons,W.G.   868,872,964,
    1100,1101,1118,1292,
    1293,1316,1317,1431,
    1451,1587,1588,1605

Maan,J.C.   300,631,794,905,
    986,987,988,1305,1306,
    1307,1308,1452,1453,1631
MacDonald,A.H.   1454,1455
MacKinnon,A.   1456
Maciel,A.C.   1652
Madhukar,A.   258,346,347,
    407,558,632,957,989,
    1092,1345,1457,1568,1569
Maekawa,S.   178,335
Mahoney,G.E.   168
Mahowald,P.   1094

Mailhiot,C.   990,991,992,
    1458,1459
Maki,K.   1460
Maki,P.A.   1461,1510
Makita,Y.   91,234,249,940
Makiuchi,M.   1635
Malik,R.J.   444,445,633,
    993,994,995,996,1156,
    1352,1353,1354,1374,1462
Maloney,T.J.   997
Manchon,D.D.,Jr.   441
Manuel,P.   136
Margalit,S.   1265,1266,
    1267,1571
Margues,G.E.   1148
Markiewicz,R.S.   1463
Marra,W.C.   348
Mars,D.E.   1274,1275,1276
Martin,R.J.   253,1283
Martin,R.M.   621
Maruyama,S.   211,245
Marzin,J.Y.   895,1464
Maslov,V.N.   185
Masselink,W.T.   1398
Massies,J.   137,173,186,
    246,349,350,351,446,447,
    448,634,848,852,998,999,
    1000,1001,1002,1003,
    1087,1204,1205,1465
Masson,J.M.   388,748,749,
    895
Masu,K.   1004,1005
Matsubara,K.   1466
Matsumoto,N.   449
Matsunaga,N.   101,187,247,
    248
Matsuoka,T.   1193
Matsushima,Y.   91,128,138,
    139,213,234,249,285
Matsuura,M.   1562
Matsuura,N.   947,948,1388
Mattord,T.J.   655,1309
Mavroides,J.G.   1472
Mayer,J.W.   46,1649
Mazur,A.   625,975,1526
McAfee,S.R.   635,1006
McCall,S.L.   324,325,326,
    327,476,893,894,1116,
    1327
McCombe,B.D.   1077
McCoy,G.L.   311
McDonald,M.L.   1518,1519,
    1520
McFee,J.H.   156,188,252,
    253,1283

McGill,T.C.   204,366,688,
  990,991,992,1458,1459
McLevige,W.V.   250,1007,
  1588
McMenamin,J.C.   295
McWhan,D.B.   419
Meeks,E.L.   226,312,921,
  1364
Meggitt,B.T.   251,450
Meillerat,C.   748
Mel'tser,B.Y.   511,645
Melchert,F.   1219,1220,1229
Melchior,H.   786
Mendez,E.E.   451,544,548,
  576,636,777,778,813,
  1008,1009,1208,1209,
  1467,1468,1469,1470,
  1489,1598,1630
Menigaux,L.   302
Merlin,R.   405,452,453,1010
Merritt,F.R.   1434
Merz,J.L.   102,140,189,190,
  218
Metze,G.M.   454,455,507,
  629,637,1471,1472
Michel,J.C.   1552
Mikkelsen,J.C.,Jr.   781
Mikulyak,R.M.   115,154
Milano,R.A.   522,1011,1473
Milanovic,V.   1012,1013,
  1474,1475
Miller,B.I.   156,188,252,
  253,1283
Miller,D.A.B.   1014,1259,
  1476,1477,1478,1648
Miller,D.L.   456,508,522,
  616,638,726,765,766,945,
  1011,1015,1016,1017,
  1018,1194,1195,1351,
  1423,1473,1479
Miller,N.J.   840
Miller,P.A.   860,861,1605
Miller,R.C.   113,141,237,
  352,377,457,639,640,641,
  642,734,901,1019,1020,
  1021,1022,1131,1480,
  1481,1482
Milnes,A.G.   1436,1437,
  1438,1439
Mimura,T.   429,458,591,592,
  594,595,596,602,603,643,
  644,744,745,914,916,917,
  918,1023,1173,1174,1483,
  1484,1485,1486
Minchev,G.M.   511,645

Misawa,S.   385,1167,1168,
  1658,1659
Mishima,T.   1004,1005
Mishimoto,C.K.   296
Misugi,T.   1546
Mitsui,S.   255,282,368
Mitsui,Y.   282
Miura,N.   956
Miura,S.   1632,1635
Miyao,M.   402,1024
Miyauchi,E.   1203,1597
Mizushima,Y.   1654
Mo,D.   1094,1302
Moench,W.   459,891,1040
Moloney,J.V.   893,1327
Mon,K.K.   646,1025
Monnom,G.   1513
Morgan,D.V.   647,648,649,
  658,1026,1157
Mori,S.290,460,461,758,1488
Moriizumi,T.   34
Morimoto,K.   254
Morkoc,H.   333,353,452,453,
  462,463,464,504,510,560,
  565,566,567,568,569,570,
  597,598,610,611,650,660,
  697,739,783,838,854,860,
  861,862,863,864,865,866,
  867,868,869,870,871,872,
  873,922,923,924,954,955,
  961,962,963,964,982,
  1007,1008,1027,1028,
  1029,1030,1031,1032,
  1033,1034,1035,1036,
  1037,1082,1100,1101,
  1118,1119,1120,1292,
  1293,1294,1295,1309,
  1314,1315,1316,1317,
  1318,1397,1398,1428,
  1429,1430,1431,1451,
  1470,1494,1573,1587,
  1588,1589,1602,1605
Mukai,S.   91,139,234,249
Mukherji,D.   103,104,142,
  143
Munoz-Yague,A.   651,768,
  769,1038,1198,1199
Munteanu,O.   639,1019,1020,
  1021,1022
Murayama,Y.   105
Muro,K.   1039,1488
Murotani,T.   255,368
Murschall,R.   1040
Muschwitz,C. v.   338
Muto,S.   960,1403

Nag, B.R.   103, 104, 142, 143,
    392, 532, 1210
Nagai, H.   608, 762, 763, 764,
    950, 951, 1193, 1390, 1391
Nagai, K.   952
Naganuma, M.   47, 74, 101, 106,
    107, 108, 109, 147, 187, 937,
    1041
Nagashima, Y.   34
Nagata, S.   191, 192
Nahory, R.E.   814
Nakagawa, T.   1487
Nakai, K.   958, 1635
Nakamura, S.   1412
Nakao, K.   354
Nakatani, M.   282, 368
Nanbu, K.   178, 335, 429, 430,
    458, 592, 593, 595, 596, 652,
    653, 931, 960, 1039, 1042,
    1375, 1403, 1488, 1547
Narayanamurti, V.   355, 1582
Narita, S.   652, 653, 1039,
    1042, 1488
Nash, F.R.   665
Nathan, M.I.   545, 546, 911,
    1043, 1489
Naughton, M.J.   1617
Neave, J.H.   238, 256, 257,
    342, 465, 624, 789, 855, 856,
    908, 975, 1140, 1417, 1490,
    1491, 1492, 1493, 1628
Nefatti, T.   1204
Nesterikhin, Y.E.   395
Neu, G.   1277
Neumann, D.A.   1494
Newman, N.   1301
Newman, P.G.   616, 638
Ng, W.   272
Nicholas, R.J.   1072, 1232,
    1495, 1496
Nicollian, E.H.   221, 222, 305
Niehaus, W.C.   167, 314
Nightingale, M.P.   1122
Niigaki, M.   402, 1024
Niina, T.   1497
Nijs, M. den   1122
Nilsson, B.   1184
Nilsson, P.O.   1594
Nishi, H.   1498
Nishi, S.   1548
Nishiuchi, K.   1173, 1486
Niwa, K.   1166
Nogami, M.   213
Noguchi, Y.   951, 1193
Nordland, W.A.   113, 141, 457

Noreika, A.J.   654, 1044,
    1045, 1499, 1646
Norris, M.T.   356, 420, 466,
    467
Norrman, S.H.   1046
Norton, N.   1491
Nottenburg, R.   1500
Nucho, R.N.   258, 346, 347
Nurmikko, A.V.   907, 1103,
    1591

O'Clock, G.D., Jr.   655, 1047
O'Connell, R.F.   1048
O'Connor, P.   820, 1049, 1515
Oberstar, J.D.   560
Obloh, H.   1401
Oe, K.   468, 656
Oelhafen, P.   1501
Ogawa, K.   1261
Ogawa, S.   211, 245
Ogura, M.   1050, 1112, 1502
Ohashi, T.   1646
Ohkawa, S.   958
Ohno, H.   390, 408, 469, 470,
    528, 647, 648, 649, 657, 658,
    685, 742, 925, 956, 1051,
    1052, 1053, 1089, 1107,
    1371, 1548
Ohta, K.   1487
Okamoto, H.   293, 340, 341,
    520, 521, 601, 607, 933, 934,
    935, 937, 1041, 1108, 1117,
    1445, 1503, 1593, 1601, 1615
Okamoto, K.   659, 1054, 1055
Olego, D.   564, 746, 1056,
    1057, 1058, 1504, 1505
Olmstead, M.   376
Olsson, N.A.   1608, 1611, 1612
Omori, M.   660, 1034
Ono, Y.   1059, 1060, 1061, 1506
Orlov, L.K.   49, 193, 265
Osaka, Y.   383, 384
Osbourn, G.C.   1322, 1507,
    1508, 1521, 1555, 1671
Ossart, P.   748
Ostapovskiy, L.M.   395
Osterling, L.E.   1356
Otsubo, M.   368
Otsuka, K.   1379, 1402
Ovchinnikov, A.A.   784

Paalanen, M.A.   1062, 1248,
    1509
Paine, B.M.   1151

Palmateer,S.C.  1163,1461,
    1510,1511
Palmberg,P.W.  1294,1309
Palmier,J.F.  471,661,832,
    1063
Palmstrom,C.  1577
Pan,C.K.  1272
Pan,S.  1094,1301,1302
Pang,Y.M.  1260,1261,1262,
    1263,1512
Panish,M.B.  20,28,94,194,
    472,473,474,475
Pankey,T.  2,7
Paparoditis,C.  1513
Paquet,D.  1233
Paradan,H.  1142
Parayanthal,P.  1577
Parayenthal,P.  1514
Park,R.M.  662,663
Parker,C.D.  1572
Parker,E.H.C.  170,251,329,
    330,450,664,967,1064
Parrinello,M.  1606
Passner,A.  324,325,326,
    327,476,893,894,1116,
    1327,1382
Patterson,G.  1599
Pattison,J.E.  230
Pavlovich,V.V.  144,145,
    195,208
Pawlik,J.R.  665,1191,1596
Pearsall,T.P.  820,828,
    1049,1072,1176,1495,
    1496,1515,1540,1603
Pearson,G.L.  323
Peck,D.D.  1572
People,R.  1516
Perry,C.H.  1123,1124,1652
Petersen,W.C.  766,1015
Petroff,P.M.  129,196,259,
    357,666,667,733,901,
    1065,1066,1177
Pettit,G.D.  612,613,743,
    1489,1501,1649
Phelps,D.E.  1517
Phillips,B.F.  1310
Phillips,J.C.  668
Phillips,J.M.  1328,1518,
    1519,1520
Picraux,S.T.  1521
Pinczuk,A.  268,269,358,
    359,477,478,479,669,700,
    1010,1056,1057,1058,
    1067,1123,1124,1504,
    1505,1522,1559,1672

Piqueras,J.  651,670,1199
Platzman,P.M.  890,1620
Pletschen,W.  723
Plew,L.  250
Pleyer,H.  513,514
Ploog,K.  197,198,216,236,
    244,260,261,262,263,289,
    317,360,361,362,363,384,
    438,439,480,509,564,618,
    619,620,671,672,673,674,
    675,676,746,858,878,886,
    940,943,944,968,969,970,
    1068,1069,1070,1071,
    1169,1170,1171,1220,
    1298,1299,1333,1383,
    1384,1409,1410,1446,
    1453,1523,1524,1535,
    1536,1666
Pokatilov,E.P.  1525
Polasko,K.J.  436
Pollak,F.H.  451,636,1514,
    1577
Pollmann,J.  337,481,625,
    975,1526
Portal,J.C.  880,1072,1232,
    1495,1496
Post,G.  749
Prange,R.E.  677,1073,1637
Price,G.L.  1074
Price,P.J.  48,678,679,680,
    1075,1076,1527,1528
Prior,K.A.  1355
Probst,C.  878,1299
Pruisken,A.M.M.  1435
Pukite,P.R.  1368,1369
Purohit,R.K.  1387

Qin,G.  1529
Queisser,H.J.  940
Quillec,M.  895
Quinn,J.J.  846,1077,1330,
    1331,1334,1529,1530,
    1613,1614

Radice,C.  167,403,462
Raisch,F.  198
Raisin,C.  388
Rakshit,S.  1531
Rammal,R.  1532
Ranke,W.  127,338,437,681,
    1533,1534
Rao,E.V.K.  572,895,1464
Rathbun,L.  470,657,658,
    737,742,1054,1055,1157,
    1190

Raymond,F.   535
Read,M.H.   1619
Regreny,A.   884,971,1211,
   1233,1416,1552
Regreny,O.   1211
Rehm,W.   1535,1536
Reich,R.K.   929,1078,1079,
   1080,1537
Reinhart,F.K.   29,51,59,
   199,500,1065,1132,1133
Rendell,R.W.   682
Reynolds,D.C.   311
Rhodin,T.N.   1549,1550
Rice,T.M.   1455
Rideau,A.   1513
Rideout,V.L.   32
Ridley,B.K.   1081,1538
Ritchie,S.   1366
Ro,C.S.   1514
Roberts,J.S.   367,394,486,
   540,683,684,691,790
Robertson,D.S.   418
Robinson,G.Y.   1622,1623
Robinson,J.W.   264
Rocher,A.   769
Rochette,J.F.   852,1003,
   1539
Rocket,A.   1082
Rode,D.L.   200
Romanov,Y.A.   25,49,93,131,
   193,265
Rosin,H.   75
Ross,R.L.   444,445,633
Rowe,J.E.   828,1540
Rowe,W.   1269,1570
Ruden,P.   564,746,943,1524,
   1535,1536,1541,1542,
   1543,1544,1545
Rupp,L.W.   96
Ruth,R.P.   508
Ryabchenko,V.E.   395
Ryuzan,O.   430,1546

Sadof'ev,Y.G.   859
Saget,P.   343
Sai-Halasz,G.A.   110,136,
   201,202,266,267,268,269,
   270,308,319,364,433,482,
   549
Saito,J.   592,603,936,1498,
   1547
Sakai,Y.   106
Sakaki,H.   164,202,270,365,
   483,685,761,925,956,
   1052,1053,1083,1084,
   1089,1107,1114,1115,1548
Sakamoto,T.   952,1085
Saku,T.   937,1379,1380,1402
Sakurai,T.   430,889,1102,
   1144,1164,1165,1325,
   1632,1633,1634,1635,1655
Salashchenko,N.N.   584
Salmon,L.G.   1549,1550
Salvi,M.   884
Samuel,G.S.   183
Sanada,T.   1144,1633,1634,
   1635,1655
Sanchez-Dehesa,J.   1086
Sang,H.W.,Jr.   1212,1213
Sapriel,J.   1551,1552
Sarin,R.   1531
Saris,F.W.   484,834,1284,
   1285
Sasa,S.   936
Sasaki,A.   1113
Sasamoto,K.   1336
Satpathy,S.   686
Saunier,P.   1553
Sauvage,M.   1087
Sauvage-Simkin,M.   1465
Savage,A.   129,259,357
Savage,R.O.   445
Saxena,R.R.   997
Scavennec,A.   749
Schaff,W.J.   648,1162,1188,
   1189,1190,1511,1653
Schaffer,W.J.   965,1151,
   1554
Schawarz,R.   422,585
Schirber,J.E.   1555
Schlapp,W.   379,729,1219,
   1220
Schlesinger,Z.   1583
Schlier,A.R.   1619
Schlosser,W.O.   167
Schmid,P.   786
Schmidt,R.R.   1556
Schneider,M.V.   203,241,
   242,271,831,1557
Schorr,A.J.   485
Schroeder,W.E.   58
Schubert,E.F.   969
Schuberth,E.   1299
Schul,G.   40
Schulman,J.N.   204,366,687,
   688,815,1253,1458,1558
Schulz,M.   70,75,1171
Schumaker,N.E.   200
Schwartz,B.   237,717
Schwartz,G.P.   689

Scifres,D.R.   205,301
Scott,G.B.   367,486,571,
   683,684,690,691,856,1088
Sebenne,C.A.   824
Segmueller,A.   54,118,206,
   306,547
Seidenbusch,W.   1440
Seki,M.   1503
Sekiguchi,Y.   685,1089,1107
Senichkina,R.S.   859
Serrano,C.M.   397,692,972
Shah,J.   207,1559,1560
Sham,L.J.   735,1148
Shank,C.V.   1090,1560
Sharma,B.D.   98
Sharma,B.L.   1387
Sharonova,L.V.   859
Shellan,J.B.   272
Shen,J.L.   1413
Shen,L.Y.L.   273
Shen,Y.D.   323
Sheng,T.T.   1619
Shibatomi,A.   958,1173,1486
Shibukawa,A.   1104
Shichijo,H.   333,421,611
Shih,H.D.   1214,1215,1553,
   1561
Shik,A.Y.   35,50,76,859
Shimanoe,T.   255,282,368
Shinozuka,Y.   1562
Shmartsev,Y.V.   22
Shmelev,G.M.   146,161,208,
   274,543,1091,1563,1564
Shon,C.M.   146
Shon,N.H.   1563
Shur,M.S.   416,431,487,488,
   574,871,872,1292,1315,
   1398,1428,1429,1430,
   1431,1623
Sibbett,W.   1282
Sigg,H.   986
Silberg,E.   814,1056,1504,
   1565,1566
Simin,G.S.   24
Simondet,F.   343
Sindalovskii,V.Y.   24
Singer,K.   1155,1156,1646
Singer,P.H.   1567
Singh,J.   1092,1568,1569
Skeath,P.   693,694,1094
Skromme,B.J.   1447
Slater,N.J.   1110,1111,1570
Smit,L.   1140,1337,1628
Smith,D.L.   1459
Smith,J.S.   1266,1267,1571

Smith,P.   1189
Smith,P.W.   1014,1477,
   1478
Smith,R.S.   425,426,489,
   615,1093
Smrcka,L.   1586
Snell,W.W.   831
Sokoloff,J.   1463
Sollner,T.C.L.G.   1572
Solomon,P.M.   1573
Somekh,S.   79,94
Son,N.H.   1564
Sovero,E.A.   1473
Spector,H.N.   1426,1427,
   1574
Spencer,M.   369
Spicer,W.E.   323,693,694,
   1094,1301,1302
Srobar,F.   695,1575
Staehli,J.L.   787
Stafeev,V.I.   26
Stagg,J.P.   790,1095
Stahl,B.   1219
Stall,R.A.   369,370,410,
   416,431,455,490,526,574,
   696,701,1576
Stamberg,R.   1037
Stanchak,C.M.   464,504,697
Stangl,G.   920
Stanley,C.R.   356,662,663,
   702,883,1514,1577,1647
Stein,D.   1578
Stern,F.   757,1528,1579
Stillman,G.E.   982,983,1447
Stoermer,H.L.   228,275,315,
   355,358,359,371,372,411,
   412,477,478,479,491,492,
   502,669,698,699,700,854,
   1096,1097,1124,1135,
   1136,1179,1248,1559,
   1580,1581,1582,1583,
   1584,1585,1617
Stokowsk,S.E.   21
Stolz,H.J.   564,746,943
Stoneham,E.B.   1599
Streda,P.   1098,1099,1586
Streetman,B.G.   250,333,
   421,464,560,565,567,610,
   611,838,954
Streifer,W.   205
Stringfellow,G.B.   701
Strom,U.   1530
Sturge,M.D.   371,372,909,
   910
Su,C.Y.   323,693,694

Su,S.L.   570,871,872,873,
    961,962,963,964,1036,
    1100,1101,1119,1120,
    1292,1431,1451,1587,
    1588,1589,1605
Su,Z.B.   1413
Suemune,I.   1656
Suffczynski,M.   1590
Suga,T.   586
Sugahara,T.   889,1102,1325
Sugai,S.   907,1103,1591
Sugimura,A.   1592
Sugiura,H.   1104
Sugiyama,K.   468,656,1105
Sugiyama,Y.   1112
Sullivan,P.W.   580,702,883,
    1106
Sun,D.C.   685,1107
Sun,Y.L.   1602
Suris,R.A.   23,33,44,276,
    784
Suyama,K.   1174
Suzuki,E.   952
Suzuki,H.   322,586
Suzuki,T.   247,373
Suzuki,Y.   285,601,1041,
    1108,1593
Svensson,S.P.   515,516,752,
    753,973,1046,1184,1185,
    1594,1595
Swaminathan,V.   665,703,
    720,1109,1191,1596

Tabatabaie-Alavi,K.   1110,
    1111,1269,1570
Tacano,M.   1112
Tai,K.   893,894,1116,1327
Takada,S.   385
Takagi,T.   1466
Takahashi,K.   34,36,47,74,
    101,106,107,108,109,147,
    187,247,248,373,1004,
    1005
Takahashi,T.   952
Takamori,A.   1203,1597
Takaoka,H.   811,1468,1598
Takase,T.   381,382
Takeda,Y.   1113
Takei,W.J.   1044,1045,1499
Takeyama,S.   652,653,1042
Talalaeff,G.   971,1211,1416
Tan,T.S.   1599
Tanaka,A.   685
Tanaka,T.   191,192
Tang,J.Y.   1374

Taniguchi,M.   685
Tannenwald,P.E.   1572
Tanoue,T.   1089,1107,1114,
    1115
Tao,R.   1600
Tarng,S.S.   893,894,939,
    1116,1327
Tarucha,S.   933,934,935,
    1117,1601
Tateishi,K.   109,147
Tausendfreund,B.   1401
Taylor,J.A.   704
Taylor,P.L.   1357
Tejayadi,O.   1587,1588,
    1589,1602
Tejedor,C.   1086
Temkin,H.   1178,1373,1603,
    1604
Tetervov,A.P.   391,985
Thiel,F.A.   156
Thornber,K.K.   841,1433
Thorne,R.E.   864,867,868,
    871,873,962,964,1036,
    1100,1101,1118,1119,
    1120,1293,1314,1316,
    1318,1451,1470,1605
Thouless,D.J.   705,1121,
    1122,1600
Tiberio,R.C.   629
Tien,P.K.   253
Tien,Z.J.   1123,1124,1652
Ting,D.Z.Y.   1254
Tiwari,S.   983
Tjapkin,D.   1012,1013,1475
Toda,K.   948
Toda,T.   1497
Todd,C.J.   428
Toledano,J.C.   1552
Tomizawa,M.   747
Tosatti,E.   1606
Toulouse,G.   1532
Tracy,J.C.   51
Trew,R.J.   1444
Tripathi,V.K.   1217
Trommer,R.   363
Trugman,S.A.   1607
Tsang,W.T.   209,210,277,
    278,279,280,374,375,376,
    377,404,485,492,493,494,
    495,496,497,498,499,500,
    539,635,639,640,641,665,
    703,706,707,708,709,710,
    711,712,713,714,715,716,
    717,718,719,720,800,801,
    802,803,804,901,978,

Tsang,W.T.
  1006,1014,1019,1020,
  1022,1109,1125,1126,
  1127,1128,1129,1130,
  1131,1132,1133,1150,
  1172,1191,1241,1242,
  1243,1244,1296,1433,
  1434,1482,1596,1608,
  1609,1610,1611,1612
Tselis,A.  1334,1613,1614
Tsu,R.  12,13,27,37,46,52,
  54,55,77,89,110,111,112,
  148,201
Tsubaki,K.  1445,1615
Tsui,D.C.  422,501,502,527,
  585,721,722,879,880,987,
  1062,1097,1134,1135,
  1136,1234,1248,1305,
  1306,1307,1308,1509,
  1581,1583,1584,1585,
  1616,1617
Tsukerman,V.G.  395
Tsurkan,G.I.  1563
Tu,C.W.  1618,1619
Tung,P.N.  849,852,980,
  1137,1138,1286
Turner,G.W.  320
Tzoar,N.  1620

Ueda,R.  1166
Uemura,C.  1104
Uihlein,C.  879,988,1306,
  1307,1308,1453
Urgell,J.J.  227,299,393,
  533
Urien,P.  1621
Urisu,T.  1654
Ury,I.  281

Vacher,R.  1552
Vaidyanathan,K.V.  250
Valeri,S.  513,835
Valois,A.J.  1622,1623
Vapaille,A.  1204
Vasquez,R.P.  1624,1625,
  1626
Vassell,M.O.  1627
Vechten,J.A. van  1139
Veen,J.F. van der  625,974,
  975,1140,1284,1285,1417,
  1418,1492,1493,1628
Venables,J.A.  149,150,180,
  183
Venkatesan,T.N.C.  324,325,
  326,327,476,1116

Verbeek,B.H.  625
Verie,C.  535
Veuhoff,E.  723
Vieren,J.P.  423,503,631,
  905,1631
Vinter,B.  1170,1629,1666
Vitlina,R.Z.  1141
Vlcek,J.  1110,1111
Vliet,C.M. van  1556
Vodjdani,N.  1142
Voisin,P.  423,503,631,724,
  905,1630,1631
Vojak,B.A.  922,923
Voos,M.  423,503,631,724,
  725,905,1143,1630,1631
Vukusic,J.I.  1282


Wada,O.  889,1102,1144,
  1164,1165,1325,1632,
  1633,1634,1635,1655
Wagner,R.J.  1234
Wagner,W.R.  200,553,554,
  555,1109,1145,1177,1178,
  1603
Waho,T.  211,245
Waldrop,J.R.  616,726,1339
Wallis,R.H.  1636
Wang,C.S.  1637
Wang,S.Y.  1146,1638
Wang,T.  873,961,1036
Wang,W.I.  606,727,728,737,
  1147,1356,1365,1424,
  1639,1640
Ward,I.  1510
Warnecke,P.  1219,1220,1229
Watanabe,H.  254
Wataze,M.  282
Watson,E.A.  893
Wecht,K.W.  1516
Weimann,G.  378,379,729,
  836,1219,1220,1298,1299,
  1578
Weinberger,D.A.  894,1327
Weisbuch,C.  377,642,667,
  730,731,732,733,734,854,
  910,1090,1560
Welch,B.M.  1300
Welch,D.  1577
Welch,D.F.  1641,1642
Welter,J.M.  266
Wemple,S.H.  928
White,S.R.  735,1148
Whiteaway,J.E.A.  1643
Whitehouse,S.B.  736

Wicks,G.W.    658,737,1149,
    1155,1370,1371,1461,
    1514,1577,1641,1642,1647
Widom,A.    1463,1644
Wiegmann,W.    51,60,87,88,
    99,102,113,129,133,141,
    190,207,228,259,275,315,
    324,325,326,327,355,357,
    358,359,371,372,389,412,
    419,422,476,477,478,479,
    502,585,667,669,698,699,
    700,722,731,732,733,734,
    894,899,900,901,909,910,
    920,939,1057,1058,1065,
    1066,1096,1116,1124,
    1259,1352,1353,1354,
    1363,1382,1478,1505,
    1559,1581,1582,1583,
    1648,1652,1672
Williams,E.R.    1234
Williams,G.F.    802,803,804,
    1150,1244
Williams,G.M.    231,418,579,
    580,843,883,1645
Williams,R.S.    666,738,1151
Williamson,W.J.    212
Witkowski,L.    611
Witkowski,L.C.    464,504,
    697,739
Wolfe,C.M.    983
Wolford,D.J.    613
Wood,C.E.C.    151,283,284,
    369,370,380,390,408,410,
    416,431,444,445,454,455,
    469,470,490,505,506,507,
    526,528,529,534,540,542,
    574,606,629,633,637,647,
    648,649,654,657,658,659,
    696,727,728,737,740,741,
    742,770,771,772,805,847,
    947,1010,1026,1044,1045,
    1054,1055,1152,1153,
    1154,1155,1156,1157,
    1162,1163,1190,1201,
    1202,1365,1370,1577,
    1646,1647,1653
Wood,T.H.    1648
Woodall,J.M.    612,613,743,
    1489,1501,1649
Woodard,D.W.    416,431,454,
    574,606,629,637,1190,
    1642
Woodbridge,K.    1158
Woodcock,J.M.    380,424,
    1350,1650,1651

Worlock,J.M.    358,359,477,
    478,479,669,1067,1123,
    1124,1159,1522,1652
Wortman,J.J.    938,1415
Wright,S.L.    436,1160,1161
Wu,E.N.    1668,1669,1670
Wuenstel,K.    970
Wyder,P.    300,794,986

Xin,S.H.    1162,1163,1439,
    1653

Yamada,S.    1654
Yamaguchi,M.    1104
Yamakoshi,S.    889,1164,
    1165,1325,1635,1655
Yamamoto,A.    1104
Yamamoto,K.    1388
Yamamoto,T.    902
Yamanishi,M.    1656
Yang,H.T.    508,1017
Yang,J.J.J.    508
Yano,M.    213,285,381,382
Yao,T.    1050,1502,1657
Yariv,A.    92,169,272,286,
    1265,1266,1267,1571
Yata,M.    1166
Yeats,R.    1246
Yeh,P.    169,272,286
Yen,R.    1560
Ying,S.C.    1331
Yokoyama,K.    747
Yokoyama,N.    744,745,1174
Yokoyama,S.    383,384
Yoneda,K.    1497
Yoshida,S.    385,1167,1168,
    1658,1659
Yoshino,J.    1548
Yoshioka,D.    1660,1661,1662
Young,I.M.    418,579,1645
Yu,L.    1413
Yu,P.W.    406
Yu,P.Y.    268,269
Yuan,H.T.    1007,1588
Yuen,S.Y.    1663
Yukitomo,K.    383,384
Yuzbashyan,E.S.    432

Zabel,H.    1494
Zaluzny,M.    1664
Zavada,J.M.    938
Zawadzki,W.    1420,1665
Zehr,S.W.    508,1017,1018
Zeller,C.    509,1169,1170,
    1666

Zheng,X.Q.   1171
Zhou,B.L.   1171
Zhu,Q.G.   966
Ziegler,J.F.   46
Ziel,A. van der   1295,1667, 1668,1669,1670
Ziel,J.P. van der   113,114, 115,152,153,154,214,215, 287,288,1172
Zilko,J.L.   1109

Zipperian,T.E.   1323,1324, 1671
Zirath,H.   831,1557
Zotos,X.   1460
Zucca,R.   522
Zucker,J.E.   1672
Zuleeg,R.   155
Zurakowski,M.P.   1301
Zurcher,P.   1213,1673

# Springer Series in Solid-State Sciences

Editors: M. Cardona, P. Fulde, H.-J. Queisser

Volume 49
**Electronic Excitations and Interaction Processes**
Proceedings of the International Symposium on Organic Materials at Schloß Elmau, Bavaria, June 5-10, 1983
Editors: P. Reineker, H. Haken, H. C. Wolf
1983. 113 figures. IX, 285 pages
ISBN 3-540-12834-3

Volume 48
**Magnetic Phase Transitions**
Proceedings of a Workshop at the Ettore Majorana Centre, Erice, Italy, July 5-13, 1983
Editors: M. Ausloos, R. J. Elliott
1983. 103 figures. VII, 269 pages
ISBN 3-540-12842-5

Volume 47
**Statics and Dynamics of Nonlinear Systems**
Proceedings of a Workshop at the Ettore Majorana Centre, Erice, Italy, July 1-11, 1983
Editors: G. Benedek, H. Bilz, R. Zeyher
1983. 117 figures. VIII, 311 pages
ISBN 3-540-12841-7

Volume 46
**Topological Disorder in Condensed Matter**
Proceedings of the Fifth Taniguchi International Symposium, Shimoda, Japan, November 2-5, 1982
Editors: F. Yonezawa, T. Ninomiya
1983. 158 figures. XII, 253 pages
ISBN 3-540-12663-5

Volume 44
I. A. Kunin
**Elastic Media with Microstructure II**
Three-Dimensional Models
1983. 20 figures. VIII, 272 pages
ISBN 3-540-12078-5

Volume 41
H. L. Skriver
**The LMTO Method**
Muffin-Tin Orbitals and Electronic Structure
1984. 34 figures. IX, 281 pages
ISBN 3-540-11519-6

Volume 40
K. Seeger
**Semiconductor Physics**
An Introduction
2nd corrected and updated edition. 1982.
288 figures. XII, 462 pages
ISBN 3-540-11421-1

Volume 39
**Anderson Localization**
Proceedings of the Fourth Taniguchi International Symposium, Sanda-shi, Japan, November 3-8, 1981
Editors: Y. Nagaoka, H. Fukuyama
1982. 98 figures. XII, 225 pages
ISBN 3-540-11518-8

Volume 38
**Physics of Intercalation Compounds**
Proceedings of an International Conference, Trieste, Italy, July 6-10, 1981
Editors: L. Pietronero, E. Tosatti
1981. 167 figures. IX, 323 pages
ISBN 3-540-11283-9

Volume 35
J. Bourgoin, M. Lannoo
**Point Defects in Semiconductors II**
Experimental Aspects
With a Foreword by G. D. Watkins
1983. 116 figures. XVI, 295 pages
ISBN 3-540-11515-3

Volume 34
P. Brüesch
**Phonons: Theory and Experiments I**
Lattice Dynamics and Models of Interatomic Forces
1982. 92 figures. XII, 261 pages
ISBN 3-540-11306-1

Volume 32
R. M. White
**Quantum Theory of Magnetism**
2nd corrected and updated edition. 1983.
113 figures. XI, 282 pages
ISBN 3-540-11462-9
(Originally published by McGraw-Hill, Inc., New York, 1970)

Springer-Verlag Berlin Heidelberg New York Tokyo

# Springer Tracts in Modern Physics

Editor: G. Höhler
Associate Editor: E. A. Niekisch

Springer-Verlag
Berlin
Heidelberg
New York
Tokyo

Volume 101
**Neutron Scattering and Muon Spin Rotation**
With contributions by numerous experts
1983. 118 figures. IX, 229 pages. ISBN 3-540-12458-6

Volume 100
**Quarks and Nuclear Forces**
Editors: B. Zeitnitz, D. C. Fries
1982. 69 figures. XI, 223 pages. ISBN 3-540-11717-2

Volume 99
**Dynamical Properties of IV-VI Compounds**
With contributions by numerous experts
1983. 47 figures. VIII, 101 pages. ISBN 3-540-12092-0

Volume 98
**Narrow-Gap Semiconductors**
With contributions by numerous experts
1983. 244 figures. X, 309 pages. ISBN 3-540-12091-2

Volume 97
W. Ehrfeld
**Elements of Flow and Diffusion Processes in Separation Nozzles**
1983. 72 figures. VIII, 140 pages. ISBN 3-540-11924-8

Volume 96
S. Büttgenbach
**Hyperfine Structure in 4d- and 5d-Shell Atoms**
1982. 14 figures. VIII, 97 pages. ISBN 3-540-11740-7

Volume 95
H. Grabert
**Projection Operator Techniques in Nonequilibrium Statistical Mechanics**
1982. 4 figures. X, 164 pages. ISBN 3-540-11635-4

Volume 94
**Exciton Dynamics in Molecular Crystals and Aggregates**
With contributions by numerous experts
1982. 37 figures. IX, 226 pages. ISBN 3-540-11318-5

Volume 93
B. Dorner
**Coherent Inelastic Neutron Scattering in Lattice Dynamics**
1982. 47 figures. VIII, 96 pages. ISBN 3-540-11049-6